OCR Revise Chemistry

A2

Exclusively endorsed by OCR for GCE Chemistry A

Second
Edition

Mike Wooster and Helen Eccles
Series editor: Rob Ritchie

www.heinemann.co.uk

✓ Free online support
✓ Useful weblinks
✓ 24 hour online ordering

01865 888080

In Exclusive Partnership

Heinemann is an imprint of Pearson Education Limited, a company incorporated in England and Wales, having its registered office at Edinburgh Gate, Harlow, Essex, CM20 2JE. Registered company number: 872828

www.heinemann.co.uk

Heinemann is a registered trademark of Pearson Education Limited

Text © Pearson Education Limited 2009

First published 2001
This edition 2009

13 12 11 10 09
10 9 8 7 6 5 4 3 2

British Library Cataloguing in Publication Data
A catalogue record for this book is available from the British Library

ISBN 978 0 435583 74 3

Edited by Sue Clements
Designed by Wearset Ltd, Boldon, Tyne and Wear
Project managed and typeset by Wearset Ltd, Boldon, Tyne and Wear
Illustrated by Wearset Ltd, Boldon, Tyne and Wear
Cover photo of the growth of fluorapatite (calcium fluoride phosphate) crystal from a hexagonal rod to sphere; composite coloured scanning electron micrographs, ©Science Photo Library
Printed in China (CTPS/02)

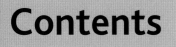

Contents

Introduction

How to use this revision guide

This revision guide is for the OCR Chemistry A2 course.

In the A2 year you will complete two written units and take two written exams called 'Rings, polymers and analysis' and 'Equilibria, energetics and elements'. Each of the two units is conveniently broken up into modules and these are listed on the page of contents. Each unit draws on what you have already learned, so you can't forget it just because you've taken the exam.

Remember to review each module more than once. This will increase your recall of the facts and the quality of your understanding.

You may find it useful to obtain a copy of the subject specification and tick off the assessable learning outcomes when you are confident that you have mastered them.

Like the AS revision guide, this book is organised into double-page spreads, which cover different sections of the specification.

At the end of each double-page spread you will see a section of 'Quick check questions'. These will test and strengthen your understanding of the basic facts for that section. The answers to these short questions are at the end of the book so you can check how well you are doing. There are also 'End-of-unit questions' that are more like exam questions; you can use these when you are near the end of your revision.

OCR A2 Chemistry

What will you study to get this qualification?

As with the AS qualification, in the second, A2, year you will study three units, and there are no optional topics. One of the three units is a practical unit. This means that you will take two written exams, called 'Rings, polymers and analysis' and 'Equilibria, energetics and elements'. There is a synoptic element to the A2 year of study in that you will use a lot of the chemical concepts you have learned at AS level and apply them in the learning of new concepts. In particular, in the second unit, the module on transition elements has a lot of links with the other units and gives you a lot of opportunity to apply your knowledge from other areas.

In addition to the two written units there is a third, practical unit. You will have to carry out three tasks from a selection provided by OCR:

- a qualitative task [10 marks]
- a quantitative task [15 marks]
- an evaluative task [15 marks].

You will perform these tasks under controlled conditions, and your teacher will assess you using a mark scheme provided by OCR. You may do more than one of each type of task, and the best mark from each will make up the overall mark out of 40.

Good luck with your revision!

Significant figures in calculations

Sig. figs, as they are known, can be important in chemical calculations. How do you work out the number of sig. figs in a number? The general rule is that *all digits are significant, except zeros that are used to position the decimal point*. Here are some rules to help you apply this generalisation.

- All **non-zero digits** are significant. So 45.9 has three sig. figs; 18.224 has five sig. figs.
- **Zeros between sig. figs** are significant. So 601.4 has four sig. figs; 100.05 has five sig. figs.
- **Zeros to the left** of the first non-zero digit are **not** significant. So 0.074 has two sig. figs; 0.0201 has three sig. figs.
- A **zero that ends a number** and is to the right of the decimal point is significant. So 43.790 has five sig. figs; 3.020 has four sig. figs.
- **Exponentials** are used to indicate the number of sig. figs where a zero ends the number and is to the left of the decimal point. So 5.300×10^3 has four sig. figs; 5.3×10^3 has two sig. figs.

To work out how many sig. figs you should use in your answer, look at the figures given in the question. You should use the same number of sig. figs in your answer as the *lowest* number of sig. figs given in the question. When you are doing the calculation, use one or two more sig. figs than this, and then round off your final answer to the correct number of sig. figs.

UNIT 1 (F324)

Rings, polymers and analysis

This is the first unit of the A2 course, and it contains many examples of chemistry from the real world.

The module contents are listed on the next page, along with reminders of the relevant work you have already completed. Different people have different revision styles, so find out what works for you and stick to it. For the first two modules you may find it helpful to concentrate on learning the reaction mechanisms first; after that it should be easier to remember the reactions of different classes of compounds. For the third module, Analysis, practise assigning peaks in spectra and answering questions.

Module 1 – Rings, acids and amines, pages 2–16

This is a module based on carbon chemistry and builds on the work you have already done at AS level in 'Chains, energy and resources'.

Double-page spread	Specification	Previous knowledge that you will use in this module
Arenes	4.1.1	Covalent bonding – σ and π bonding.
Electrophilic substitution	4.1.1	Electrophiles and 'curly arrows'. Revise mechanisms.
Phenol	4.1.1	You will use your knowledge of the arenes here.
Carbonyl compounds	4.1.2	Oxidation and reduction. Review work on alcohols.
Carboxylic acids	4.1.3	Look at acids and the esterification reaction of alcohols from AS work.
Esters	4.1.3	Review esterification.
Nitrogen compounds	4.1.4	The reactions of halogenoalkanes with ammonia.
Phenylamine ($C_6N_5NH_2$) and azo compounds	4.1.4	This is new, but remember balancing of equations.

Module 2 – Polymers and synthesis, pages 17–27

This module looks closely at synthesis and the chemistry of larger molecules. You need to have mastered the material from the previous module because this module utilises the principles learned there.

Double-page spread	Specification	Previous knowledge that you will use in this module
Amino acids	4.2.1	You will need to have mastered the chemistry of the –COOH and –NH_2 groups. Remember the tetrahedral shape around the central carbon in a molecule.
Amino acids, proteins and peptides	4.2.1	This is new material but you will need to use your knowledge of pH and what it means in terms of the concentration of hydrogen ions.
Stereoisomerism	4.2.1	Revise *E/Z* isomerism from the 'chains' section at AS level.
Polymerisation	4.2.2	Revise addition polymerisation.
Condensation polymerisation	4.2.2	The work on polypeptides and proteins is relevant here.
Organic synthesis including chiral synthesis	4.2.2	This incorporates a lot of what has gone before and is a good opportunity to revise all the reactions you have come across.

Module 3 – Analysis, pages 28–37

You will use all your chemistry here as you see how analytical techniques can be used both to separate and to identify compounds. This is where it all comes together and makes sense.

Double-page spread	Specification	Previous knowledge that you will use in this module
Chromatography	4.3.1	General vocabulary and mass spectrometry.
Carbon-13 NMR	4.3.2	Isomerism and working out the isomers from a molecular formula are very important here.
Proton NMR	4.3.2	Isomerism and working out the isomers from a molecular formula are very important here.
Identifying compounds using NMR spectroscopy	4.3.2	Isomerism and working out the isomers from a molecular formula are very important here.
Combined analytical techniques	4.3.2	You will bring together a lot of the work you have done in all three modules here.

End-of-unit questions, pages 38–39

UNIT 1 Arenes

Key words

- arene
- delocalised
- σ-and π-electrons
- benzene ring
- substitution
- isomerism

✔*Quick check 1*

Arene chemistry and aliphatic (straight chain) chemistry make up the two main branches of organic chemistry. 'Arene' is the ' Rings' part of 'Rings, acids and amines'! Arene chemistry is the chemistry of the benzene ring.

Background facts

The molecular formula of benzene is C_6H_6 so its empirical formula is **CH**. You can spot an arene derivative from the relative numbers of carbon and hydrogen atoms. If the numbers are almost equal or there are more carbons than hydrogens then the compound is an arene; for example, methylbenzene is C_7H_8 and naphthalene is $C_{10}H_8$.

The structure of benzene

Using skeletal formula there are two ways of representing the structure of benzene. These are called the Kekulé structure (named after the scientist who discovered it) and the **delocalised** structure.

In the Kekulé structure the π-electrons in the three double bonds are isolated and therefore benzene could be considered a triene. This would make it very reactive.

In the delocalised structure, the **π-electrons** in the three double bonds are spread around the whole **benzene ring**.

The evidence from the reactions of benzene and studies of its structure strongly supports the delocalised structure as being the actual structure of benzene.

The evidence in favour of the delocalised system compared with the Kekulé structure is summarised below:

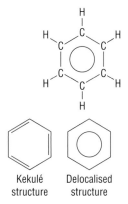

Kekulé structure Delocalised structure

Experimental evidence	How does this support the delocalised structure?	How does it show that the Kekulé structure is incorrect?
X-ray studies show the carbon–carbon bond lengths are all the same, being somewhere in between the length of a C—C bond and a C=C bond.	The delocalised system means that the bonds are all the same type (in between double and single) and therefore the same length.	In the Kekulé structure there are two types of bond (C—C and C=C) and therefore there would be two bond lengths.
The $\Delta H_{hydrogenation}$ of benzene is -208 kJ mol^{-1} not -360 kJ mol^{-1}. Benzene is therefore more stable than expected.	The delocalisation of the π-electrons would stabilise the structure and lower the energy released when it is hydrogenated.	The $\Delta H_{hydrogenation}$ of cyclohexene is -120 kJ mol^{-1}. The Kekulé structure would give an enthalpy change of 3×-120 kJ mol^{-1} (-360 kJ mol^{-1}). Cyclohexene
Benzene does not decolorise (does not react with) bromine water. Benzene needs a catalyst (halogen carrier) for the reaction.	The delocalised π-electron system has insufficient electron density to polarise the Br–Br bond and react. A catalyst aids the formation of a Br$^+$ ion and this is sufficiently positive to react with the delocalised electrons.	The three isolated double bonds would have sufficient electron density to polarise the Br–Br bond and react. No catalyst would be required.

Therefore the displayed formula of benzene is as shown here.

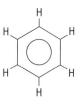

All the bond angles are 120°. Benzene's structure is a very important aspect of its chemistry and should be understood before you learn its reactions. How benzene's structure comes about is shown on the right.

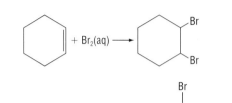

The reactivity of benzene compared with alkenes

Because it is a cyclic compound, the alkene cyclohexene (C_6H_{10}) is a good comparison. It will react with dilute bromine water in the cold; no catalyst is required. This is an addition reaction.

To react with bromine, benzene requires a catalyst, and the bromine is not diluted. This is a **substitution**.

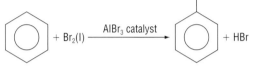

Isomerism and naming in benzene compounds

Because of the ring structure, benzene compounds can exhibit positional **isomerism.** For example, there are three isomers of dichlorobenzene, as shown here:

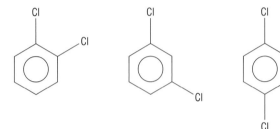

| 1,2-dichlorobenzene | 1,3-dichlorobenzene | 1,4-dichlorobenzene |

If the substituents are different and can bond together, then more than three isomers may be obtained. For example there are four compounds of formula C_7H_7Cl.

To name them, the carbon atoms in the benzene ring are numbered 1 to 6. The place of substitution is always the lowest possible.

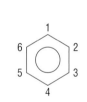

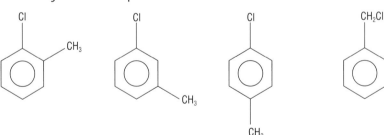

1-chloro-2-methylbenzene   1-chloro-3-methylbenzene   1-chloro-4-methylbenzene   chloromethylbenzene

✓ Quick check 2

✓ Quick check 3 and 4

✓ Quick check 5

QUICK CHECK QUESTIONS

1 Which of the following molecular formulae could represent arene compounds?
(a) C_7H_8 (b) C_6H_{14}
(c) $C_8H_{10}O$ (d) $C_4H_{10}O$.

2 (a) Give three pieces of evidence that support the delocalised structural model of benzene rather than the Kekulé structure.
(b) Using electron orbitals, explain how the delocalised structure comes about.

3 Draw the structures of the following compounds:
(a) 1,3-dibromobenzene   (b) 1,2-dimethylbenzene
(c) 1,3,5-trichlorobenzene   (d) 1-bromo-3-ethylbenzene
(e) 1-chloro-4-methylbenzene.

4 Draw the four isomers of the arene with the molecular formula C_8H_{10}.

5 Name the following compounds:
(a) ⬡— Cl   (b) H_3C —⬡— CH_3
(c) ⬡— $CH_2CH_2CH_3$

Electrophilic substitution

Key words

- electrophiles
- electrophilic substitution
- halogen carriers

✓ *Quick check 1*

✓ *Quick check 2*

- Substitution allows the benzene to retain its delocalised system of π-electrons and hence its stability.
- The high electron density above and below the plane of the benzene ring means that the ring is liable to attack from an electron-deficient atom or group.
- Such groups are called **electrophiles**. *Electrophiles are electron-pair acceptors.* The substitution of a hydrogen on the ring by another atom or group of atoms is called **electrophilic substitution.**
- You have to know nitration and halogenation as examples.

Nitration

The mononitration of benzene uses concentrated nitric acid and concentrated sulfuric acid at 50 °C. This is one of the two mechanisms you should learn.

- The full equation for the reaction is:

$$C_6H_6 + HNO_3 \rightarrow C_6H_5NO_2 + H_2O$$

- The concentrated sulfuric acid acts as a catalyst in this reaction, and the electrophile in the reaction is the *nitronium (NO_2^+) ion* (or nitryl cation).

This mechanism is best followed as a number of steps:

STEP 1 The formation of the electrophile – the nitronium ion (or nitryl cation).

We can summarise the reaction as:

$$H_2SO_4 + HNO_3 \rightarrow HSO_4^- + NO_2^+ + H_2O$$

Examiner tip

Remember, a curly arrow represents the movement of a pair of electrons.

STEP 2 The NO_2^+ ion is a powerful electrophile and attracts electrons out of the ring to form a positively charged intermediate. This intermediate has only four π-electrons (compared with six for benzene) and the electrons are delocalised over only five carbon atoms.

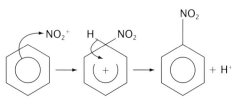

STEP 3 The electrons in the positively charged intermediate are only partially delocalised, making it unstable. To regain a completely delocalised system a hydrogen ion (proton) is lost to give nitrobenzene. The proton then reacts with the HSO_4^- ion to regenerate the sulfuric acid, which is the catalyst.

$$H^+ + HSO_4^- \rightarrow H_2SO_4$$

Halogenation

Halogen atoms replace hydrogen atoms on the ring. **Halogen carriers** such as iron, iron(III) bromide or aluminium chloride catalyse the reaction.

A halogen carrier is an electron-deficient molecule that helps produce an electron-deficient halogen atom or alkyl group (both electrophiles). This can then react with the electrons on the benzene ring.

If the halogen is chlorine then the equation for the reaction is:

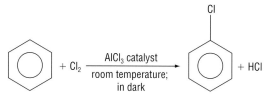

✓ *Quick check 3*

STEP 1 The formation of the electrophile (a positive halogen ion). This is heterolytic fission of the Cl_2 molecule into a Cl^+ and a Cl^- ion (the latter forming a dative covalent bond with the $AlCl_3$).

STEPS 2 and 3 are similar to those for mononitration. At the end, the catalyst, $AlCl_3$, is regenerated by reaction of the proton formed with $AlCl_4^-$.

STEP 2

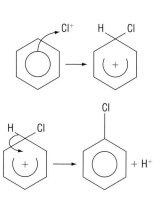

STEP 3

$$H^+ + AlCl_4^- \rightarrow AlCl_3 + HCl$$

Summary of electrophilic substitution reactions

Except for the generation of the electrophile these reactions are very similar. The electrophile can be represented as E^+ and the final product as C_6H_5E. This is shown in the diagram below:

✓ *Quick check 3*

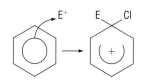 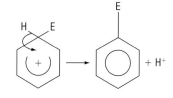

QUICK CHECK QUESTIONS

1 Explain why benzene undergoes substitution reactions and not addition reactions.

2 Write the balanced equation for the nitration of benzene and write out the mechanism, explaining each step.

3 Complete the following equations and for each one describe the conditions required for the reaction.
   (a) $C_6H_6 + Cl_2 \rightarrow A$ (organic molecule) + B
   (b) $C_6H_6 + C \rightarrow D + C_6H_5Br$
   (c) $C_6H_6 + HNO_3 \rightarrow E$ (organic molecule) + F

Phenol (C_6H_5OH)

The structure of **phenol** is shown opposite.

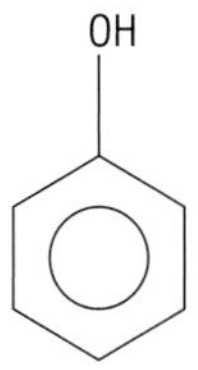

This molecule can be represented by its formula C_6H_5OH.

Because it has a hydroxyl (OH) group attached to a benzene ring, its chemistry can be treated in two parts:
1 That of the OH group
2 That of the benzene ring.

However, each one affects the other, which makes phenol an interesting molecule to study.

Key words

- phenol
- proton donor
- acid

Reactions of the –OH group

- Phenol loses the proton of its –OH group more easily than alcohols because the presence of the benzene ring weakens the O–H bond. Thus phenol can act as a **proton donor** in **aqueous solution**: it is a very weak **acid**.
  Therefore if a proton acceptor or base (Base:) is present the following reaction can take place.

  C_6H_5OH + Base: $\rightarrow$ $C_6H_5O^-$ + Base : H^+

- With aqueous alkalis the phenol acts as an acid, forming water and the phenoxide ion.
  Example: With sodium hydroxide, water and sodium phenoxide are formed.

  $NaOH + C_6H_5OH \rightarrow C_6H_5O^-Na^+ + H_2O$

- With sodium, phenol gives sodium phenoxide and hydrogen gas.

  $Na + C_6H_5OH \rightarrow C_6H_5O^-Na^+ + \frac{1}{2}H_2$

✓ *Quick check 1*

Hint

The base has a lone pair of electrons which are used to form a dative covalent bond with the proton.

✓ *Quick check 2*

Activation of the benzene ring in phenols

I   The lone-pair electrons on the oxygen of the –OH group can be delocalised *on the benzene ring*.

II This increases the electron density on the ring, especially at the sites indicated by the closed circles in the diagram opposite (positions 2, 4 and 6 on the ring), and makes the benzene ring more reactive.

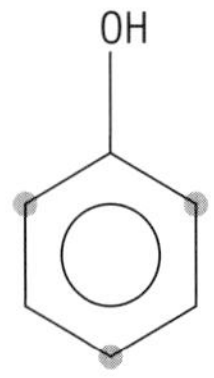

Therefore when the benzene ring of phenol reacts with electrophiles:

- *It reacts more readily than benzene* because of the increased electron density on the benzene ring. This means that the conditions needed for phenol to react are much less severe than for benzene.

- Substitution occurs at the sites indicated in the diagram above because of the increased electron density at these sites compared with the others.

- Substitution often occurs *not just at one of these sites but at all three* because of the extra reactivity.

Example: Phenol with bromine:

I   No catalyst (halogen carrier) is required.

II The reaction occurs with bromine water. Compare this with the bromination of benzene, which requires pure bromine to brominate the benzene ring. This shows the increased reactivity of the benzene ring in phenol.

III Substitution occurs at three carbons (positions 2, 4 and 6) to give 2,4,6-tribromophenol.

$$C_6H_5OH + 3Br_2 \rightarrow C_6H_2Br_3OH + 3HBr$$

2,4,6-tribromophenol

OR

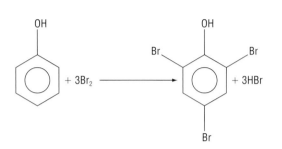

The uses of phenolic compounds

Plastics

Phenolic resins are formed through the reaction of phenols with simple aldehydes. They are used for making moulded products such as snooker and pool balls. Phenolic resins are also used along with, for example, paper to make 'paper phenolics', which have uses as electrical components. 'Glass phenolics' are suited for use in the high-speed bearing market.

Resins in paints

Phenolic resins are used in making varnishes.

Antiseptics

Compounds containing phenol groups are often used to kill microbes. Thymol and TCP (the proprietary name) are used as antibacterial mouthwashes.

Disinfectants

'Dettol' (4-chloro-3,5-dimethylphenol) is a commonly used household disinfectant. It is harmful to microbes but relatively harmless to humans.

Dettol

QUICK CHECK QUESTIONS

1   Explain why phenol is an acidic compound.

2   Complete the following equations, giving the products and balancing the equation.
   (a) $C_6H_5OH + KOH \rightarrow$
   (b) $C_6H_5OH + Na \rightarrow$
   (c) $C_6H_5OH + Cl_2 \rightarrow$
   (d) $C_6H_5OH + Li \rightarrow$

3   When phenol reacts with bromine water, substitution takes place at the 2, 4 and 6 positions on the benzene ring. Explain why this is so.

4   Give evidence to show that the –OH group in phenol activates the benzene ring.

5   Give three uses for phenol compounds.

Carbonyl compounds

Examiner tip

Carboxylic acids and esters also contain the >C=O group but have other functional groups on the same carbon atom and can therefore be distinguished from carbonyl compounds.

The **aldehydes** and **ketones** are two closely related homologous series. They both contain the **carbonyl** (>C=O) **group** and are therefore called **carbonyl compounds**.

Background facts

- The general formula of carbonyl compounds is $C_nH_{2n}O$ and, once we get to three carbons, aldehydes are isomeric with ketones. Aldehydes and ketones are **unsaturated** compounds.

- When naming aldehydes the names end in -al, e.g. ethanal and butanal.

- The names of ketones end in -one, and a number preceding it indicates where on the carbon chain the carbonyl group is situated. Two examples are butan-2-one and pentan-2-one.

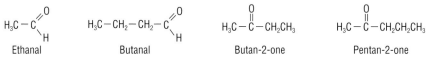

| Ethanal | Butanal | Butan-2-one | Pentan-2-one |

The preparation of aldehydes and ketones

Aldehydes are prepared by the *oxidation of primary alcohols* using acidified potassium dichromate. The oxidation has to be carefully controlled and the product distilled off immediately so that the aldehyde is not oxidised further to give a carboxylic acid.

e.g. butanal is prepared from butan-1-ol.

$$CH_3CH_2CH_2CH_2OH + [O] \rightarrow CH_3CH_2CH_2CHO + H_2O$$

   1° alcohol   Oxidising agent   Aldehyde

Ketones are prepared by the *oxidation of secondary alcohols* using acidified potassium dichromate.

e.g. butan-2-one is prepared from butan-2-ol.

$$CH_3CH_2CH(OH)CH_3 + [O] \rightarrow CH_3CH_2COCH_3 + H_2O$$

   2° alcohol   Oxidising agent   Ketone

Nucleophilic addition to carbonyl compounds

The electronegative oxygen adjacent to the carbonyl carbon means that a dipole is produced in the C=O bond. The carbon atom is therefore electron-deficient and can be attacked by **nucleophiles**.

$$\overset{\backslash}{\underset{/}{C}} = \overset{}{O} \atop {\delta+ \ \delta-}$$

Examiner tip

Nucleophiles are compounds or ions that can donate a pair of electrons. In this example it is the H^- ion. In practice $NaBH_4$ in water is used.

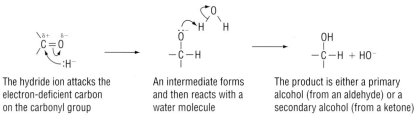

The hydride ion attacks the electron-deficient carbon on the carbonyl group

An intermediate forms and then reacts with a water molecule

The product is either a primary alcohol (from an aldehyde) or a secondary alcohol (from a ketone)

Because the compounds are unsaturated the reaction is also an addition, so the reaction mechanism is that of **nucleophilic addition.** The example shows the mechanism for the reaction with hydride (H:⁻) ions from sodium borohydride.

Generally, the reaction between any carbonyl compound and H^- ions may be written as shown above.

The reduction of carbonyl compounds

The reagent used is sodium borohydride, $NaBH_4$, in water.

Group	Example
Aldehydes are reduced to primary alcohols	Ethanal is reduced to ethanol
$RCHO + 2[H] \rightarrow RCH_2OH$	$CH_3CHO + 2[H] \rightarrow CH_3CH_2OH$
Ketones are reduced to secondary alcohols	Butan-2-one is reduced to butan-2-ol
$RCOR' + 2[H] \rightarrow RCH(OH)R'$	$CH_3COCH_2CH_3 + 2[H] \rightarrow CH_3CH(OH)CH_2CH_3$

Examiner tip

The use of [H] shows the use of a reducing agent.

Distinguishing and identifying carbonyl compounds

The presence of an aldehyde or ketone carbonyl group can be identified using a solution of **2,4-dinitrophenylhydrazine (2,4-DNP, Brady's reagent)**.

Once the presence of a carbonyl compound is confirmed, you can distinguish between aldehydes and ketones using either **Tollens' reagent** or **acidified potassium dichromate**.

Which aldehyde or ketone is present can be confirmed using the precipitate from the reaction with 2,4-dinitrophenylhydrazine.

Examiner tip

Only aldehydes and ketones form precipitates with 2,4-DNP. Other compounds containing the >C=O group, such as carboxylic acids and esters, do not.

	Aldehydes	Ketones
Identification test Add 2,4-dinitrophenylhydrazine solution. **This also confirms the presence of the aldehyde or ketone carbonyl group.**	They give an orange crystalline precipitate called a 2,4-dinitrophenylhydrazone. Each aldehyde gives a 2,4-dinitrophenylhydrazone with a characteristic melting point after it has been recrystallised, and therefore can be identified using existing data.	They give an orange crystalline precipitate called a 2,4-dinitrophenylhydrazone. Each ketone gives a 2,4-dinitrophenylhydrazone with a characteristic melting point after it has been recrystallised, and therefore can be identified.
Distinguishing test (I) Add ammoniacal silver nitrate solution, Tollens' reagent (a mild oxidising agent).	They are oxidised to a carboxylic acid and a **silver mirror** is formed. $RCHO + [O] \rightarrow RCOOH$	No reaction
Distinguishing test (II) Add acidified potassium dichromate solution.	Aldehydes are oxidised to carboxylic acid and the potassium dichromate changes colour from orange to green. $RCHO + [O] \rightarrow RCOOH$	No reaction and therefore the dichromate stays orange in colour.

✓ *Quick check 4*

QUICK CHECK QUESTIONS

1 Name the following aldehydes:
(a) $CH_3CH_2CH_2CHO$
(b) $CH_3CH(CH_3)CHO$
(c) C_6H_5CHO
(d) $C_6H_5CH_2CHO$.

2 Name the following ketones:
(a) CH_3COCH_3
(b) $CH_3CH_2COCH_2CH_3$
(c) $C_6H_5CH_2COCH_3$
(d) $CH_3COCH_2CH_2C_6H_5$.

3 Complete the following equations, giving the structure of the carbon compound formed.
(a) (i) $CH_3CHO + 2[H] \rightarrow$
(ii) $CH_3COCH_2CH_3 + 2[H] \rightarrow$
(b) Name and give the structural formulae of the products after reducing the following carbonyl compounds:
(i) $C_6H_5COCH_3$
(ii) $CH_3CH(CH_3)CHO$.

4 A student was asked to identify an unknown compound. He suspected that it was a carbonyl compound. Explain how he could confirm his suspicions and then identify the compound.

Examiner tip

The silver ions are reduced because they gain electrons:

$Ag^+ + e^- \rightarrow Ag$

Examiner tip

[O] represents the oxygen from the oxidising agent.

Carboxylic acids

✔ *Quick check 1*

Carboxylic acids are weak acids, but they are the strongest acids you will study in A-level organic chemistry. They are important because they are the starting materials for making esters and they are found naturally, making them readily available.

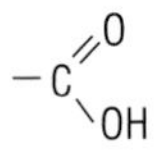

Background facts

- The functional group present is the **carboxylic acid** (–COOH) group, which is trigonal planar in shape with bond angles of 120°.
- Their general formula is $C_nH_{2n}O_2$. Their names are derived from the alkane with the same number of carbons in the longest carbon chain.
- If there are two carboxyl groups in the molecule, then again the name is derived from the parent alkane and the two carboxyl groups are indicated by the suffix -dicarboxylic acid (or -dioic acid).

Name	Molecular formula	Structural formula
Methanoic acid	CH_2O_2	HCOOH
Ethanoic acid	$C_2H_4O_2$	CH_3COOH
Butanoic acid	$C_4H_8O_2$	$CH_3CH_2CH_2COOH$
2-methylpropanoic acid	$C_4H_8O_2$	$CH_3CH(CH_3)COOH$

- Dicarboxylic acids, e.g. ethanedioic acid, $(COOH)_2$, are important in the preparation of **condensation polymers** such as **polyesters**, which are used for making bottles and fibres for clothing (see page 24).
- Short-chain carboxylic acids are soluble in water because they form hydrogen bonds with water molecules and this makes it energetically favourable for them to dissolve.

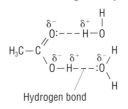

Hydrogen bond

✔ *Quick check 2*

Carboxylic acids as acids

In the carboxyl group the two electronegative oxygens pull electrons towards them, weakening the O–H bond. This makes them **proton donors**.

Example:

$$CH_3COOH(aq) + H_2O \rightleftharpoons CH_3COO^-(aq) + H_3O^+(aq)$$

In aqueous solution, carboxylic acids are weakly acidic; 1 mol dm^{-3} solution of a carboxylic acid has a pH of 2.8 or more, compared with a pH of 0 for the same concentration of strong acids such as hydrochloric acid.

Carboxylic acids react with reactive metals such as magnesium to form hydrogen gas and a salt.

Example:

$$2CH_3COOH + Mg \rightarrow (CH_3COO)_2Mg + H_2$$
magnesium ethanoate

✔ *Quick check 3*

With alkalis such as sodium hydroxide they react just like other acids in forming a salt plus water and are therefore neutralised by the alkali.

$$CH_3COOH + NaOH \rightarrow CH_3COO^-Na^+ + H_2O$$
ethanoic acid · · · · · · · · · · sodium ethanoate

✓*Quick check 3*

The ionic equation may be written:

$$CH_3COOH + OH^- \rightarrow CH_3COO^- + H_2O$$
ethanoic acid · · · · · · · · · ethanoate ion

With carbonates, carbon dioxide, water and a salt are always formed.

$$2CH_3COOH + Na_2CO_3 \rightarrow 2CH_3COO^-Na^+ + CO_2 + H_2O$$
ethanoic acid · · · · · · · · · sodium ethanoate

Esterification

In the presence of acid catalysts and heat, carboxylic acids and alcohols react to form esters and water. The reaction is a reversible one.

Carboxylic acid + alcohol $\rightleftharpoons$ ester + water

✓*Quick check 4*

- The acid catalyst may be either concentrated hydrochloric or sulfuric acid. It is the hydrogen ions supplied by the acids which are the actual catalysts.
- For example, ethanoic acid and ethanol react to form ethyl ethanoate and water.

$$CH_3COOH + CH_3CH_2OH \rightleftharpoons CH_3COOCH_2CH_3 + H_2O$$

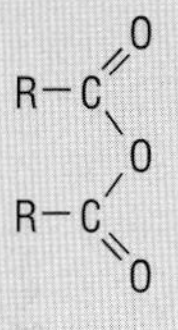

Acid anhydrides

Acid anhydrides are more reactive derivatives of carboxylic acids; they can form esters with alcohols without the need for a catalyst.

Acid anhydride + alcohol $\rightleftharpoons$ ester + carboxylic acid

e.g. ethanoic anhydride, $(CH_3CO)_2O$, reacts with ethanol to form the ester ethyl ethanoate and ethanoic acid.

$$(CH_3CO)_2O + CH_3CH_2OH \rightleftharpoons CH_3COOCH_2CH_3 + CH_3COOH$$

Examiner tip

Acid anhydrides have the structure shown below. They are more reactive than carboxylic acids. (R = CH$_3$–, C$_2$H$_5$– etc.)

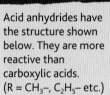

✓*Quick check 4*

QUICK CHECK QUESTIONS

1 Give the structural formulae of the following acids:
   (a) propanoic acid
   (b) 2-methylbutanoic acid
   (c) 2-phenylethanoic acid.

2 Explain why short-chain carboxylic acids dissolve in water.

3 Complete the following equations:
   (a) $CH_3CH_2COOH + NaOH \rightarrow$
   (b) $CH_3CHClCOOH + Na_2CO_3 \rightarrow$
   (c) $C_6H_5COOH + Na \rightarrow$
   (d) $CH_3CH(CH_3)COOH + Mg \rightarrow$
   (e) $C_6H_5COOH + NaOH \rightarrow$

4 Complete the following equations:
   (a) $CH_3CH_2COOH + CH_3OH \rightleftharpoons$
   (b) $(CH_3CH_2CO)_2O + CH_3OH \rightleftharpoons$
   (c) $CH_3CHClCOOH + CH_3CH_2CH_2OH \rightleftharpoons$
   (d) $(CH_3CO)_2O + CH_3CH_2CH_2OH \rightleftharpoons$
   (e) $(CH_3CO)_2O + C_6H_5CH_2OH \rightleftharpoons$

Esters

Facts you should know

- **Esters** are isomeric with carboxylic acids; therefore, their general formula is $C_nH_{2n}O_2$. They can, however, be distinguished from carboxylic acids by their infrared spectra, *which do not show* the broad absorption band at 3000–2500 cm^{-1} given by carboxylic acids.
- Their characteristic functional group is the ester bond (–CO–O–). Structurally they may be represented as RCOOR' where R and R' are carbon chains or rings, e.g. CH_{3-}.
- Esters are named according to the acid and alcohol from which they are prepared. The alcohol part of the name comes first and the acid part second. For example, ethyl ethanoate and methyl propanoate:

$$H_3C - \overset{\overset{O}{\|}}{C} - O - C_2H_5 \qquad C_2H_5 - \overset{\overset{O}{\|}}{C} - O - CH_3$$

ethyl ethanoate   methyl propanoate

Uses of esters

- In perfumes and flavourings. Many natural perfumes are esters, for example oil of wintergreen and the scents of flowers. Examples of flavourings are ethyl methanoate (raspberry essence) and pentyl ethanoate (pear essence).
- As solvents: used for dissolving such substances as drugs and antibiotics.
- In **biodiesel**, which is emerging as a renewable alternative to diesel from fossil fuels.

The formation of esters (esterification)

Esters are formed by the acid-catalysed reversible reaction of acids and alcohols.

Acid + alcohol ⇌ ester + water

$$CH_3COOH + CH_3CH_2OH \overset{H^+ \text{ catalyst}}{\rightleftharpoons} CH_3COOCH_2CH_3 + H_2O$$

Fatty acids; fats and oils

- Fats are esters of long-chain carboxylic acids (**fatty acids**) and the alcohol glycerol (propane-1,2,3-triol, $CH_2OHCHOHCH_2OH$).

$$H_2C - \underset{OH}{CH} - CH_2 \\ \underset{OH \quad OH \quad OH}{}$$

- These esters are called **triglycerides**. An example is given below:

$$H_2C - OCO(CH_2)_{16}CH_3$$
$$HC - OCO(CH_2)_7CH=CH(CH_2)_7CH_3$$
$$H_2C - OCO(CH_2)_4CH=CHCH_2CH=CH(CH_2)_7CH_3$$

- The fatty acids that form the esters with glycerol can be either saturated or unsaturated. The number of carbons and the number and placement of the double bonds are explained in the name.

Examples of these fatty acids are shown in the table below.

Name	Saturated or unsaturated	Placement of double bonds	Structure
Octadecanoic acid, 18,0. The number of carbons is shown by the 18 and the number of double bonds by the 0	Saturated since no double bonds	Not applicable	$CH_3(CH_2)_{16}COOH$
Octadec-9-enoic acid, 18,1(9)	Unsaturated with one double bond	Between the 9th and 10th carbons of the chain (starting at the carboxyl group)	$CH_3(CH_2)_7CH=CH(CH_2)_7COOH$
Octadeca-9,12-dienoic acid, 18,2(9,12)	Unsaturated with two double bonds	Between the 9th and 10th and between the 12th and 13th carbons of the chain	$CH_3(CH_2)_4CH=CHCH_2CH=CH-(CH_2)_7COOH$

- The greater the number of carbon–carbon double bonds, the lower the melting point of the fatty acid and hence of the triglyceride. ✓*Quick check 4*
- Unsaturated fatty acids make up the triglycerides found in vegetable and fish oils whilst saturated fatty acids tend to make up the triglycerides found in animal fats.

Fatty acids and our health

- *Trans* fatty acids (*trans* fats) are formed when manufacturers add hydrogen to vegetable oil (hydrogenation), and this type of fat is associated with the raising of the levels of low-density lipoprotein or 'bad cholesterol'.
- 'Bad cholesterol' is associated with an increased risk of obesity, coronary heart disease and strokes.
- The levels of *trans* fat in foods has now got to be written on food labels. ✓*Quick check 5*

Acids and alkaline hydrolysis of esters and triglycerides

- The reverse reaction is a **hydrolysis** reaction because water is split (hydro = water; lysis = splitting) as shown below. When fats are hydrolysed by alkalis (see later), salts of long-chain fatty acids (soaps) are formed and the process is called saponification.

For example, in the presence of an acid catalyst such as sulfuric acid, ethyl ethanoate is hydrolysed to ethanoic acid and ethanol.

$$CH_3COOC_2H_5 + H_2O \overset{H^+ \text{ catalyst}}{\rightleftharpoons} CH_3COOH + C_2H_5OH$$
ester　water　carboxylic acid　alcohol ✓*Quick check 6*

A fat would be hydrolysed to glycerol and the carboxylic acids combined with it.

- If an alkali (e.g. NaOH) is used, then the reaction goes to completion, producing an alcohol and a carboxylate (RCOO–) ion.

For example, in the presence of an alkali, ethyl ethanoate is hydrolysed to sodium ethanoate plus ethanol. This reaction goes to completion.

$$CH_3COOC_2H_5 + OH^- \rightarrow CH_3COO^- + C_2H_5OH$$

Transesterification is the reaction between triglycerides and another alcohol to form esters of the alcohol and glycerol. The new esters formed are used in **biodiesel.**

QUICK CHECK QUESTIONS

1 Give the structural formulae of all the structural isomers (carboxylic acids and esters) with the following molecular formulae:
 (a) $C_3H_6O_2$ (three isomers)
 (b) $C_4H_8O_2$ (five isomers).

2 (a) Give the structural formulae and the names of the carboxylic acids forming the following esters:
　 (i) $CH_3CH_2COOCH_3$
　 (ii) $HCOOCH_3$
　 (iii) $CH_3CH_2CH_2COOCH_2CH_3$
 (b) Name the esters in (i) to (iii) above.

3 Name some uses for esters and explain why you think they have these uses.

4 Give the structures of the following fatty acids:
 (a) hexadec-9-enoic acid 16,1(9)
 (b) octadeca-9,12-dienoic 18,2(9,12).

5 Explain the health concerns associated with *trans* fats.

6 Complete the following equations:
 (a) $CH_3COOC_2H_5 + H_2O \rightarrow$
 (b) $C_6H_5COOCH_3 + NaOH \rightarrow$
 (c) $CH_3CH_2COOC_6H_5 + H_2O \rightarrow$
 (d) $HCOOCH_2CH_2CH_3 + H_2O \rightarrow$

Nitrogen compounds

These form a group of very important compounds. Primary **amines** are industrially important because of their uses in making dyes.

Background facts

- They are derivatives of ammonia (NH_3).
- The functional group of primary amines is the amino ($-NH_2$) group.
- They are named according to the group to which they are attached.

Examples are given in the table below:

Structural formula	Name
CH_3NH_2	Methylamine
$CH_3CH_2NH_2$	Ethylamine
$C_6H_5NH_2$	Phenylamine

✓*Quick check 1*

The basic nature of primary amines

Examiner tip

The H^+ ions need two electrons to be stable and these are provided by a dative covalent bond with the nitrogen.

- The new $R-NH_3^+$ ion is an ammonium (NH_4^+) ion in which an alkyl or aryl group has been substituted for one of the hydrogen atoms. These new ions are named by placing the name of the alkyl or aryl group in front of the word ammonium. In the two examples given, an ethyl group (C_2H_5) or a phenyl group substitute one hydrogen atom, so the ions are named ethylammonium ion and phenylammonium ion.

$$CH_3CH_2NH_2 \ + \ H^+ \rightarrow CH_3CH_2NH_3^+$$
ethylamine   methylammonium ion

$$C_6H_5NH_2 \ + \ H^+ \rightarrow \ C_6H_5NH_3^+$$
phenylamine   phenylammonium ion

- Because they are **proton acceptors**, primary amines are Bronsted–Lowry **bases**.
- With water, they lead to the formation of OH^- ions, so they are weak alkalis.

e.g. $CH_3CH_2NH_2 + H_2O \rightleftharpoons CH_3CH_2NH_3^+ + OH^-$

✓*Quick check 2*

- Because they act as bases, primary amines react with acids to form salts. The names of these salts come from the ions formed from the amine and the acid. For example, ethylammonium chloride and phenylammonium chloride are formed by the reactions of the bases ethylamine and phenylamine with hydrochloric acid.

✓*Quick check 3*

For example:

$$NH_3 \ \ + HCl \rightarrow \ \ \ NH_4^+Cl^-$$
Ammonia   ammonium chloride

$$CH_3CH_2NH_2 + HCl \rightarrow CH_3CH_2NH_3^+Cl^-$$
Ethylamine   ethylammonium chloride

$$C_6H_5NH_2 \ \ + HCl \rightarrow \ \ C_6H_5NH_3^+ Cl^-$$
Phenylamine   phenylammonium chloride

✓*Quick check 4*

Preparation of amines

There are two methods – one for straight-chain amines and one for aromatic amines.

I Preparation of straight-chain (aliphatic) amines
- Excess ammonia is refluxed with a halogenoalkane (usually the bromoalkane) with ethanol as the solvent. You can describe this as an ethanolic solution of ammonia. The general equation for the reaction can be written as follows:

$$RBr + NH_3 \rightarrow RNH_2 + HBr \qquad [R = CH_3\text{--}; C_2H_5\text{--}; (CH_3)_2CH\text{--}; etc.]$$

The excess ammonia then reacts with the HBr to form ammonium bromide:

$$NH_3 + HBr \rightarrow NH_4^+Br^-$$

e.g.

$$CH_3CH_2Br + NH_3 \rightarrow CH_3CH_2NH_2 + HBr$$

$$NH_3 + HBr \rightarrow NH_4^+Br^-$$

II Preparation of aromatic amines
- The nitro compound is refluxed with a mixture of tin and concentrated hydrochloric acid.
- The general equation may be written as:

$$ArNO_2 + 6[H] \rightarrow ArNH_2 + 2H_2O$$

An example is shown below.

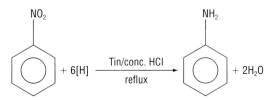

Examiner tip

Here Ar represents the benzene ring or a derivative of benzene.

Examiner tip

In practice, in the presence of hydrochloric acid, the phenylamine forms the salt, $C_6H_5NH_3^+Cl^-$. To get the free amine, NaOH solution is added.

$$C_6H_5NH_3^+Cl^- + NaOH \rightarrow C_6H_5NH_2 + NaCl + H_2O$$

QUICK CHECK QUESTIONS

1 Name the following compounds:
   (a) CH_3NH_2
   (b) $CH_3CH_2CH_2NH_2$
   (c) $CH_3CH(CH_3)CH_2NH_2$
   (d) $C_6H_5NH_2$.

2 (a) Complete the following equations:
   (i) $CH_3NH_2 + H_2O \rightleftharpoons$
   (ii) $C_6H_5NH_2 + H_2O \rightleftharpoons$
   (b) Why are the amines in (a) able to act as bases?

3 Complete the following equations and name the products:
   (a) $CH_3NH_2 + HCl \rightarrow$
   (b) $CH_3CH_2CH_2NH_2 + HCl \rightarrow$
   (c) $C_6H_5NH_2 + HCl \rightarrow$
   (d) $CH_3NH_2 + H_2SO_4 \rightarrow$
   (e) $CH_3NH_2 + CH_3COOH \rightarrow$

4 Give equations for the preparation of the following amines:

   (a) 

   from 1-methyl-4-nitrobenzene

   (b) $CH_3CH_2CH_2NH_2$

   from 1-bromopropane

   (c)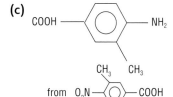

   from O_2N—⬡—COOH

   (d)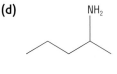

   from 2-bromopentane

UNIT 1

Phenylamine ($C_6H_5NH_2$) and azo compounds

- Phenylamine and other aromatic amines are the starting compounds for the formation of an important class of compounds – the **azo dyes**. The reaction scheme and accompanying details for this are given below.

Stage 1 Sodium nitrite ($NaNO_2$) and hydrochloric acid react to form nitrous acid. The temperature should be below 10 °C because the **benzenediazonium** chloride decomposes at higher temperatures.

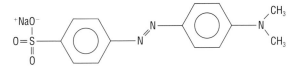

Benzenediazonium chloride

Stage 2 The benzenediazonium chloride is then added to phenol in the presence of an alkali; an azo dye is formed.

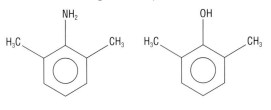

The azo group (–N=N–) is the group responsible for the colour of the dye formed.

Economically this reaction is very important since the formation of azo-dye compounds is the basis of part of the dye industry.

A common example of an azo dye is the acid–base indicator methyl orange.

QUICK CHECK QUESTIONS

1. Describe how you could synthesise an azo dye using phenylamine and phenol.

2. Give the structure of the amine and the other organic reagent that could be used to prepare methyl orange.

3. Give the structure of the azo dye that could be prepared from the following two compounds.

Amino acids

Background facts

Their general formula is RC*H(NH₂)COOH and their general displayed formula can be represented as shown opposite.

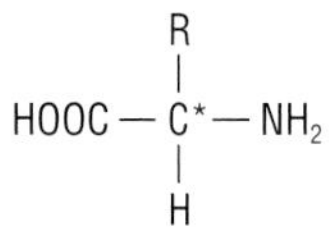

$$HOOC - \overset{\overset{\displaystyle R}{|}}{\underset{\underset{\displaystyle H}{|}}{C^*}} - NH_2$$

The asterisk (*) indicates the chiral carbon.

If R is not hydrogen, the carbon atom marked with an asterisk is attached to four different atoms/groups. It is therefore a **chiral** carbon, and amino acids can exhibit **optical isomerism**. Optical isomers can be drawn as mirror images of each other.

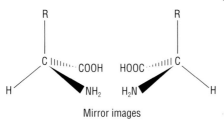

Mirror images

The acid–base nature of amino acids

All amino acids contain the following two groups:

1 The basic amine ($-NH_2$) group. This group can accept protons.
2 The carboxylic acid ($-COOH$) group. This group can donate protons.

- Because they contain these two groups, they can react with both acids and bases (they are **amphoteric**).
- As acids, the carboxyl group can donate a proton to a base to form a salt plus water.

$$RCH(NH_2)COOH + NaOH \rightarrow RCH(NH_2)COO^- Na^+ + H_2O$$

- As bases, the amino group can accept a proton from an acid to form a salt.

$$RCH(COOH)NH_2 + HCl \rightarrow RCH(COOH)NH_3^+Cl^-$$

Key words

- amino acids
- chirality
- optical isomerism
- amphoteric

Examiner tip

Together, the sodium nitrite and hydrochloric acid form nitrous acid, HNO_2.

Examiner tip

You can read more about optical isomerism on page 20.

Examiner tip

The $-NH_2$ group acts as a base by accepting a proton:

$$H^+ + -NH_2 \rightarrow -NH_3^+$$

The $-COOH$ group acts as an acid by donating a proton:

$$-COOH \rightarrow -COO^- + H^+$$

Examiner tip

When you draw optical isomers, you should try to give some idea of three dimensions to the figures.

✓*Quick check 1*

QUICK CHECK QUESTION

1 **(a)** Complete the following equations:
    (i) $CH_2(NH_2)COOH + HCl \rightarrow$
    (ii) $CH_2(NH_2)COOH + NaOH \rightarrow$

(b) Explain why the amino acid is acting as a base in reaction **(i)** and as an acid in reaction **(ii)**.

Amino acids, proteins and peptides

Key words

- amino acids
- zwitterions
- isoelectric point
- proteins
- peptide bond
- condensation polymers

✓ *Quick check 1*

✓ *Quick check 3*

Amino acids are white crystalline solids that have much higher melting points than would be expected from their molecular masses.

The explanation for these properties is that at intermediate pH values they exist as a type of 'inner salt' or **zwitterion**, where the charge on the amino group ($-NH_3^+$) cancels out that on the carboxylate group ($-COO^-$). The general form of the zwitterion of an amino acid is shown in the diagram.

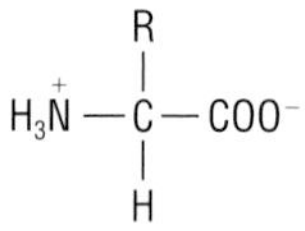

$$H_3\overset{+}{N} - \underset{\underset{H}{|}}{\overset{\overset{R}{|}}{C}} - COO^-$$

The positive and negative charges on each zwitterion attract opposite charges on neighbouring zwitterions, forming ionic bonds. These are stronger than the hydrogen bonds that would exist between the uncharged amino acids. This stronger ionic bonding explains the higher than expected melting points of amino acids.

The effect of pH on the charges on the zwitterions

- At high pH values the proton on the $-NH_3^+$ will be removed, leaving an overall negative charge.

- At low pH values the $-COO^-$ group will accept a proton, giving an overall positive charge.

- At an intermediate pH called the **isoelectric point**, there will be equal negative and positive charges and the amino acid will have no overall charge and will not move when placed in an electric field. This pH is different for each amino acid.

✓ *Quick check 2*

- For some amino acids R can be a basic group (e.g. R is $CH_2CH_2CH_2CH_2NH_2$), in which case the isoelectric point is raised to a high value (alkaline pH). For others R can be an acidic group (e.g. R is $CH_2CH_2CH_2COOH$) and the isoelectric point will be low (an acidic pH).

Proteins and peptides

Facts you should know

- **Proteins** are chains of amino acids linked by peptide bonds. The **peptide bond** is the $-CONH-$ link.

Examiner tip

The amide link is

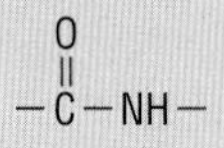

$$-\overset{\overset{O}{\|}}{C} - NH -$$

- Proteins are formed from amino acids by the *loss of water* as the peptide bonds are formed. Proteins are therefore **condensation polymers**.

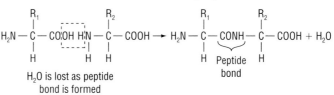

H_2O is lost as peptide bond is formed   Peptide bond

Examiner tip

A polypeptide is one containing more than ten amino acids – they usually contain 100–300.

A common exam question involves the structures of the **di**peptide formed when *two* amino acids combine. The point is that the amino acids can combine in two different ways to give compounds with *different* structures.

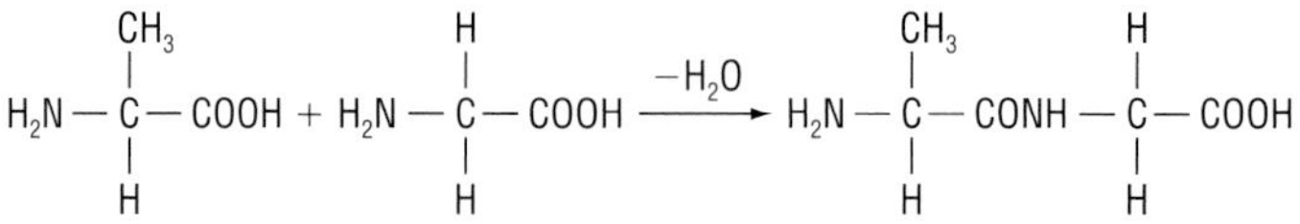

$$H_2N - \underset{\underset{H}{|}}{\overset{\overset{CH_3}{|}}{C}} - COOH + H_2N - \underset{\underset{H}{|}}{\overset{\overset{H}{|}}{C}} - COOH \xrightarrow{-H_2O} H_2N - \underset{\underset{H}{|}}{\overset{\overset{CH_3}{|}}{C}} - CONH - \underset{\underset{H}{|}}{\overset{\overset{H}{|}}{C}} - COOH$$

In this reaction the carboxyl group of 2-aminopropanoic acid reacts with the amine group of 2-aminoethanoic acid.

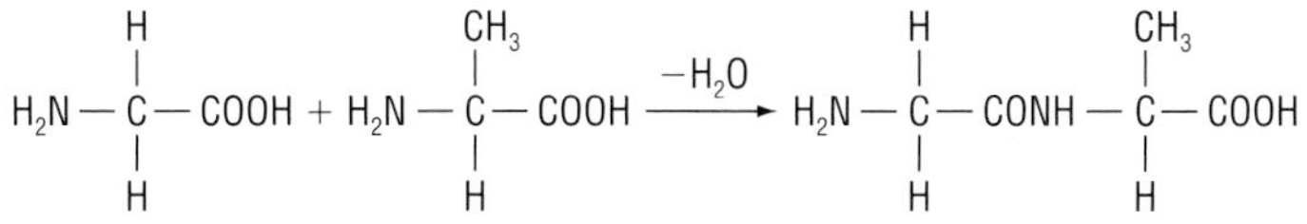

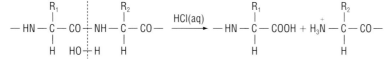

In this reaction the amine group of 2-aminopropanoic acid reacts with the carboxyl group of 2-aminoethanoic acid.

To get the individual amino acids back from a polypeptide, water has to be added back to the polypeptide and the peptide bonds broken. This is another example of hydrolysis.

* When a protein or peptide is refluxed with moderately concentrated (approximately 6 mol dm^{-3}) hydrochloric acid for several hours, it is hydrolysed and the protein is split up into its constituent amino acids. This is acid-catalysed hydrolysis.

* The same reaction occurs if the hydrolysis is catalysed using concentrated sodium hydroxide. The only difference is that the products will be carboxylates.

* Hydrolysis is essentially the reverse of the formation of the peptide bond.

The peptide bond is hydrolysed as it reacts with water. The terminal amine group, $-NH_2$, accepts a proton from the acid to become $-\overset{+}{N}H_3$.

QUICK CHECK QUESTIONS

1 Draw the zwitterions formed from the following amino acids:
  (a) $CH_2(NH_2)COOH$
  (b) $CH_3CH(NH_2)COOH$
  (c) $C_6H_5CH(NH_2)COOH$.

2 The isoelectric points of the acids shown above are as follows:

2-aminoethanoic acid
$CH_2(NH_2)COOH$
pH 5.97

2-aminopropanoic acid
$CH_3CH(NH_2)COOH$
pH 6.00

2-phenyl-2-aminoethanoic acid
$C_6H_5CH(NH_2)COOH$
pH 5.48

  (a) Draw the ions that would be formed by 2-aminoethanoic acid at:
    (i) pH 2.00
    (ii) pH 5.97
    (iii) pH 10.00.
  (b) Draw the structures of the predominant ions that would be formed by:
    (i) 2-aminopropanoic acid at pH 5
    (ii) 2-phenyl-2-aminoethanoic acid at pH 7.

3 Why are the melting points of the three amino acids in question 2 greater than would be expected from their molecular masses?

4 **(a)** Draw the *two* dipeptides that would be formed by combining the pairs of amino acids below:
    (i) $CH_2(NH_2)COOH$ and $CH_3CH(NH_2)COOH$
    (ii) $CH_2(NH_2)COOH$ and $C_6H_5CH(NH_2)COOH$
    (iii) Two molecules of $CH_2(NH_2)COOH$.
  (b) Draw the structures of the amino acids responsible for the formation of the dipeptides shown below:
    (i)
    (ii)

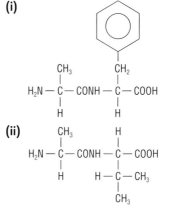

Stereoisomerism

Stereoisomers have the same structural formula, but different displayed formulae. That is, they differ by the arrangement of their atoms in space. Stereoisomerism is of paramount importance to the organic chemist, especially if the compound is to be used in living systems. There are two types of stereoisomerism – ***E/Z* isomerism** and **optical isomerism**. An important natural example of *E/Z* isomerism is the existence of *E* and *Z* isomers of retinol (vitamin A) in the mechanism of vision. In living organisms, vast numbers of enzymes and molecular receptors are able to recognise optical isomers; if the isomer is the wrong one it will not be used. For example, the smell of lemons and oranges comes from compounds with the same structural formula, limonene, but from different optical isomers of limonene.

E/Z isomerism

✓ *Quick check 1*

- This requires a C=C double bond (as found in alkenes) and two different groups on each carbon atom in the double bond.

- The atoms or groups on the C=C bond are assigned priorities. For example, a CH_3– group has a higher priority than a hydrogen atom. If the two higher priority groups are on different carbons but on the same side of the C=C bond then the compound is the *Z*-isomer. If they are diagonally opposite each other it is the *E*-isomer.

- *Cis–trans* isomerism is a special case of *E/Z* isomerism in which two of the groups, one on each C atom, are identical.

Examples are the two isomers of pent-2-ene. In this example they can also be described as *cis–trans* isomers because two of the groups attached to the C=C are identical.

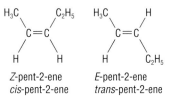

Z-pent-2-ene
cis-pent-2-ene

E-pent-2-ene
trans-pent-2-ene

Optical isomerism

- This requires a **chiral centre** or **chiral carbon** (one carbon atom *attached to four different groups or atoms*). Each **chiral** carbon atom is often indicated by an asterisk (*).

- **Optical isomers** are mirror images of each other and are drawn as such. One isomer cannot be superimposed on the other. Like our left and right hands they are *non-superimposable*.

- The property of existing as optical isomers is called **chirality**.

✓ *Quick check 2*

- The optical isomers of a chiral compound react in exactly the same way with other reagents. They do, however, show very different biological activity, and in the pharmaceutical industry this is very important (see page 26).

Examples are all the α-amino acids (apart from 2-aminoethanoic acid or glycine), and 2-hydroxypropanoic acid.

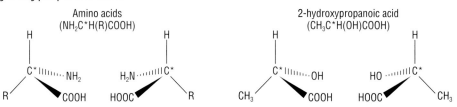

Amino acids
($NH_2C^*H(R)COOH$)

2-hydroxypropanoic acid
($CH_3C^*H(OH)COOH$)

Optical isomerism is very important in biological molecules. Optical isomers will have very different biological effects from one another. For example, limonene has two optical isomers. One smells of lemons whilst the other smells of oranges.

Distinguishing between optical isomers

The solutions or the crystals of optical isomers can be distinguished from one another because they rotate plane-polarised light. The isomer rotating the light clockwise is called the (+) isomer, whilst the other isomer is the (−) isomer.

Optical isomerism in biological molecules

Many molecules have more than one chiral carbon and in the exam you may be asked to identify chiral carbons. The diagram below shows an ester found in coconuts. The left-hand diagram shows the molecule and the right-hand diagram shows the same molecule with the chiral carbons ringed.

✔ *Quick check 3 and 4*

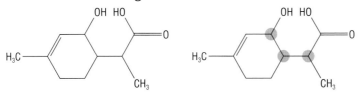

QUICK CHECK QUESTIONS

1 (a) Draw the stereoisomers of the following alkenes:
    (i) but-2-ene
    (ii) $C_6H_5CH=CHCH_3$.
   (b) Why doesn't but-1-ene exhibit *E/Z* isomerism?
   (c) Both pent-2-ene, $CH_3CH=CH(C_2H_5)$, and the compound $CH_3(C_2H_5)C=CHCl$ have *E/Z* isomers whilst only pent-2-ene has *cis–trans* isomers. Explain why.

2 Explain why when enzymes catalyse reactions they do not produce products that are a mixture of optical isomers.

3 (a) Draw the optical isomers of $CH_3CH(OH)COOH$. In your diagram indicate the chiral carbon atoms with an asterisk.
   (b) Explain why $CH_3C(CH_3)(OH)COOH$ is not optically active.
   (c) Draw the optical isomers of 2-aminopropanoic acid $CH_3CH(NH_2)COOH$ and indicate the chiral carbon using an asterisk (*).

4 Limonene has two optical isomers. Their structures are given below. Draw out the structures and ring the chiral carbons.

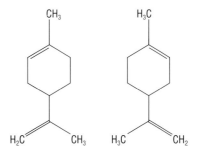

Polymerisation

Key words

- addition polymerisation
- condensation polymerisation
- monomer
- repeat unit

Examiner tip

An addition polymer is a very long molecular chain, formed by repeated addition reactions of many unsaturated alkene molecules (the monomers).

✓ *Quick check 1*

You have already studied **addition polymerisation** as part of the chemistry of the alkenes at AS level. We will review this and then go on to look at the other type of polymerisation – **condensation polymerisation**.

Review of addition polymerisation

In the AS section on addition polymerisation we looked at the polymerisation of alkenes. Remember, addition polymerisation requires an alkene and involves the opening out of the double bonds to connect the molecules together and give a polymer with all single bonds.

The diagram below of the polymerisation of polyphenylethene summarises the changes involved.

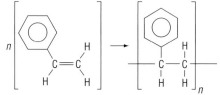

Think of the alkene as having four groups attached, W, X, Y and Z. The polymer formed from this **monomer** can be drawn as shown below.

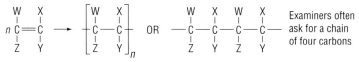

Examiners often ask for a chain of four carbons

For example, but-1-ene can be written as shown below, making it easy to draw the polymer

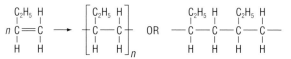

Condensation polymerisation

- A condensation reaction may be defined as one in which two molecules join together to form a larger molecule with the elimination of a smaller molecule. An example of a condensation reaction is esterification:

$$CH_3COOH + C_2H_5OH \rightarrow CH_3COOC_2H_5 + H_2O$$

larger molecule   small water molecule eliminated

- In condensation polymerisation the requirements are that the monomer or monomers making up the polymer have 2 functional groups. These functional groups can be identical (as in the general example below) or different (as in lactic acid, see page 25).

Examples of the groups X and Y are
—OH, —COOH and —NH₂

- The reaction still leaves X and Y available for bonding and the reaction can continue to give a polymer. In this general case:

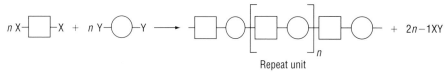

Repeat unit

- The lines represent bonds between the units and these are usually either the ester (–CO–O–) bond or the amide (–CO–NH–) bond.

Examples are shown in the table below:

No.	X—[ ]—X	Y—◯—Y	Molecule XY	Bond formed, type of polymer and repeat unit (shown by brackets)
1	Ethane-1,2-diol $HOCH_2CH_2OH$	Benzene-1,4-dicarboxylic acid 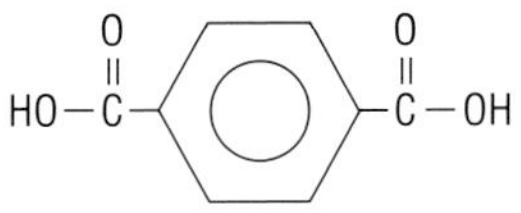 HO—C(=O)—◯—C(=O)—OH	H_2O	Ester (—CO—O—), polyester 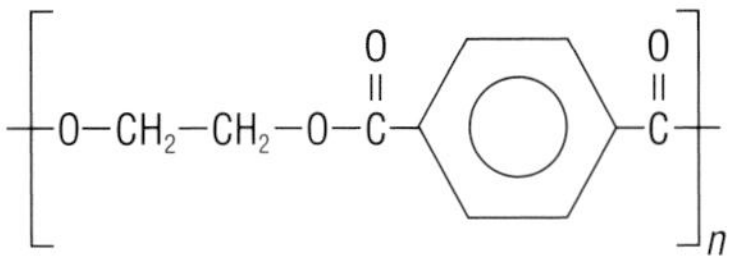 [—O—CH_2—CH_2—O—C(=O)—◯—C(=O)—]_n
2	1,6-diaminohexane $H_2N(CH_2)_6NH_2$	Hexane-1,6-dicarboxylic acid $HOOC(CH_2)_4COOH$	H_2O	Amide (CO—NH—), polyamide 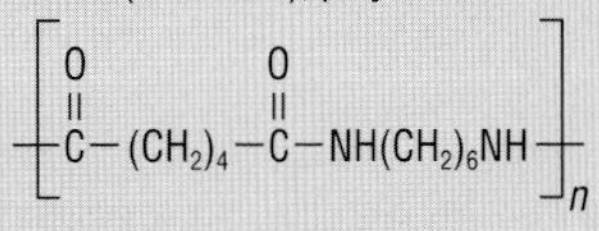 [—C(=O)—(CH_2)_4—C(=O)—NH(CH_2)_6NH—]_n Nylon 6,6
3	Benzene-1,4-diamine 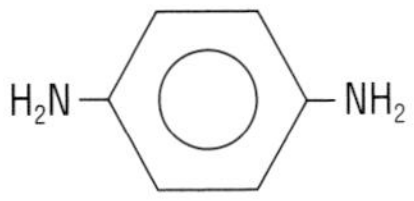 H_2N—◯—NH_2	Benzene-1,4-dicarboxylic acid 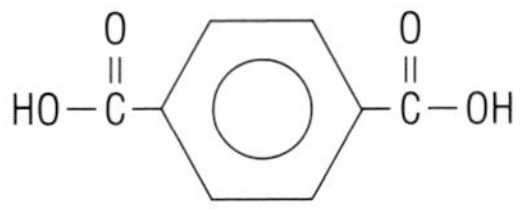 HO—C(=O)—◯—C(=O)—OH	H_2O	Amide (CO—NH—), polyamide 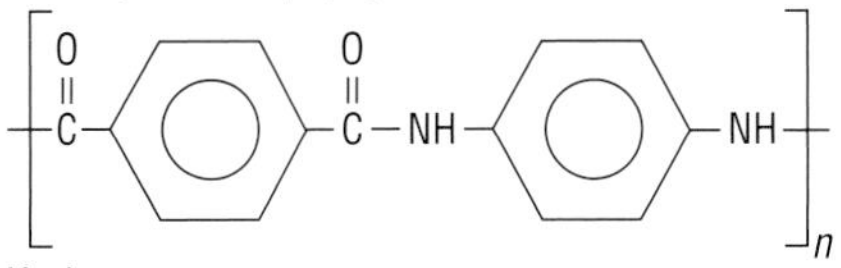 [—C(=O)—◯—C(=O)—NH—◯—NH—]_n Kevlar
4	Amino acid 1 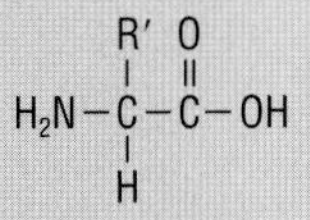 H_2N—C(R')(H)—C(=O)—OH	Amino acid 2 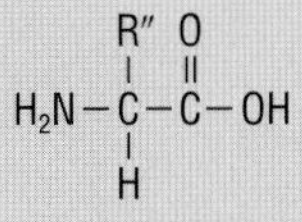 H_2N—C(R'')(H)—C(=O)—OH	H_2O	Amide (peptide) bond (CO—NH—), polypeptide 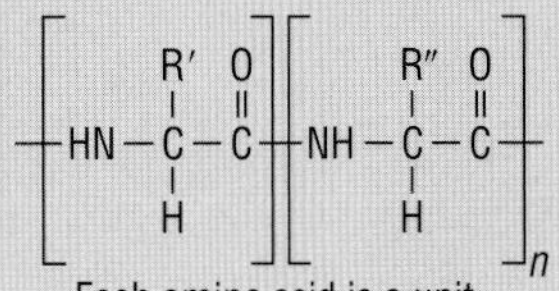 [—HN—C(R')(H)—C(=O)—][—NH—C(R'')(H)—C(=O)—]_n Each amino acid is a unit

Module 2

The uses of polyesters and polyamides in fibres

✔ *Quick check 2*

- The strong bonds in polyamides and polyesters make the polymer chains themselves strong, and ideal for making fibres.
- The long chains of polyamides in nylon fibres are held together by hydrogen bonds, as shown in the diagram below. These hydrogen bonds can be easily broken and reformed, making nylon fibres elastic. In Kevlar (used for bullet-proof vests), the perfect alignment of the chains, and multiple hydrogen bonding, make it a very strong material.

✔ *Quick check 3*

- For polyesters such as Terylene, cross-linking occurs through dipole–dipole bonding. Again, the fibres produced are elastic and therefore good for making clothing.

QUICK CHECK QUESTIONS

1 Draw the polymers formed from the monomer units **(a)** to **(d)** below.

(a) 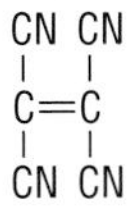 CN CN
 C=C
 CN CN

(b) CH_3 H
 C=C
 CH_3 H

(c) H Cl
 C=C
 H H

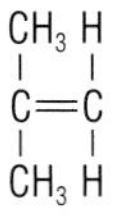

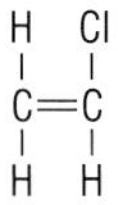

(d) 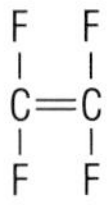 F F
 C=C
 F F

2 Ethane-1,2-diol ($HOCH_2CH_2OH$) can form condensation polymers. Draw the repeating units of the polymers it forms with the compounds shown below.

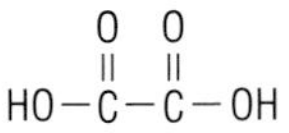

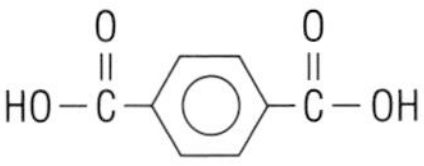

HO—C(=O)—C(=O)—OH   HO—C(=O)—◯—C(=O)—OH

3 **(a)** Explain why nylon is elastic.
   (b) Why do polyesters and polyamides make good fibres?

UNIT 1 Condensation polymerisation – monomers, repeat units and uses of condensation polymers

Module 2

Key words

- hydrolysis
- recycling
- photodegradability
- biodegradability

A common task set by examiners is the identification of repeat units and the monomers that are their components. This section is designed to give you some idea of how to tackle this type of question.

Deducing the repeat units and monomers in condensation polymers

A typical condensation polymer is shown below.

To determine the repeat unit for the polymer, go through the following steps:

STEP 1 Draw a bracket line, as shown, where a distinctive group begins or ends. In this case we have selected between the carbonyl and amine groups.

STEP 2 Follow the chain until you come to just before this is about to repeat itself and draw the other bracket.

✓Quick check 1

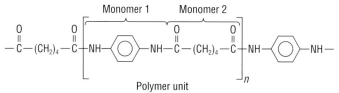

To determine the monomer units go through the following steps:

STEP 1 Determine what the linkage is in the polymer. In this case it is the amide (–CO–NH–) linkage.

STEP 2 What are the two monomer units on either side of the linkage? To make an amide linkage one monomer must be a diamine and the other a dicarboxylic acid. In this case we have the following:

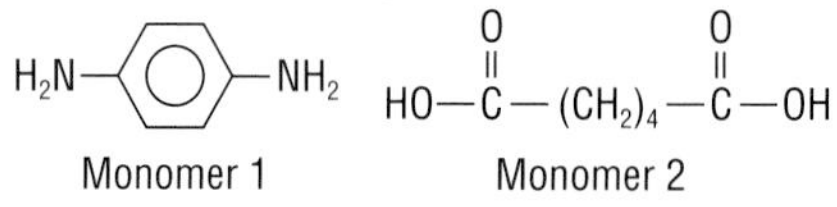

✓Quick check 2

Similarly, if the polymer is a polyester, then the linkage is the ester (–CO–O–) linkage. An example is the polyester shown below:

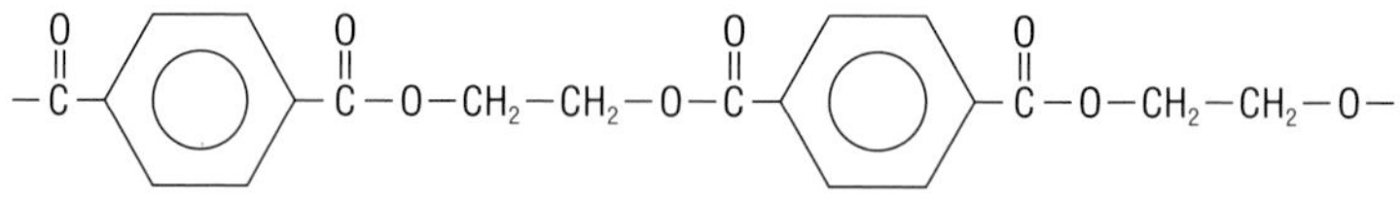

The hydrolysis of condensation polymers

Both polyesters and polyamides are hydrolysed in the presence of acids and alkalis. Examples are:

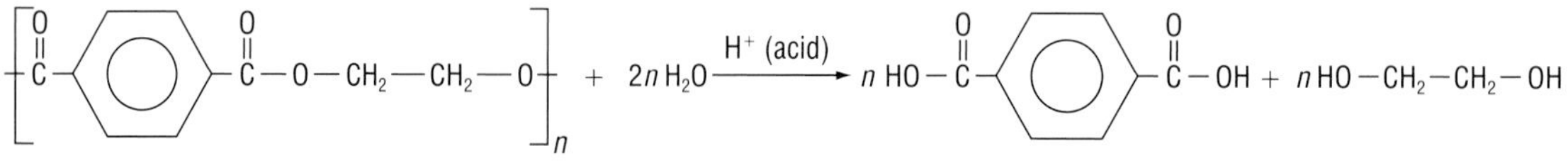

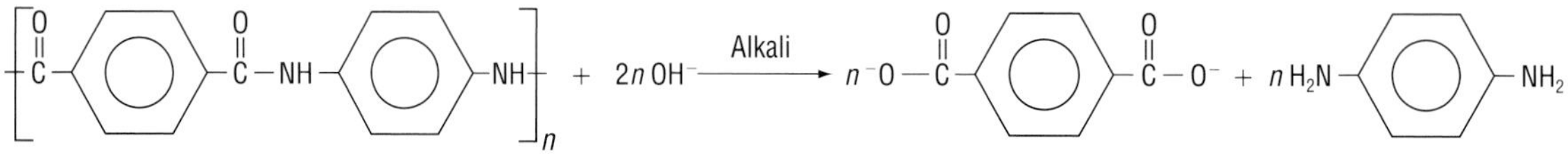

Problems with polymers

- Many polymers are not **biodegradable** and therefore persist in the environment, becoming a litter problem as well as a danger to wildlife. For this reason chemists are trying to produce polymers that are degradable, thus minimising environmental waste.

Disposal of polymers by burning is problematic because toxic fumes are produced unless very high temperatures are used.

 Quick check 3

How can environmental waste be minimised?

The C=O bond absorbs radiation. Incorporating these bonds into the polymer chain will mean that, when radiation is absorbed by the bond, the energy absorbed causes breakdown of the chain. This is called photodegradability.

The ester and amide bonds can be hydrolysed by acids and alkalis, thus helping their breakdown and disposal.

Another way round the disposal problem is to produce polymers formed from plant feedstock such as corn starch, wheat and sugar beet. The advantages and disadvantages of using polymers based on starch are shown in the table.

Advantages	Disadvantages
Polymers made from plant feedstock are biodegradable – microbes transform them into H_2O and CO_2 and other harmless substances.	Land used to grow corn starch for polymers is not used for growing food.
They are carbon neutral.	The shelf-life of packaging is reduced.
Litter degrades more quickly, reducing the litter problem.	If mixed with other plastics for recycling, their value is reduced.
They do not use fossil fuels. They are formed from renewable sources.	They have a poorer mechanical strength than plastics based on fossil fuels.

Poly(lactic acid) (PLA)

Lactic acid (2-hydroxypropanoic acid) has two functional groups (secondary alcohol and carboxyl group) and can therefore be made into a polyester.

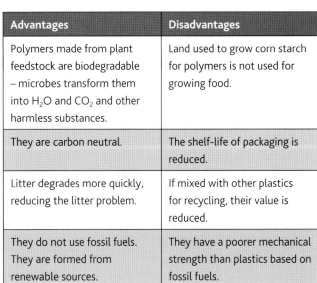

- Lactic acid is a sustainable feedstock for this reaction because it can be extracted from corn starch and sugar cane.

 Module 2

 Quick check 4

- It is a chiral compound and therefore has optical isomers. One form, L-lactic acid, is the naturally occurring form made by bacterial fermentation. It can be polymerised to form poly-(L-lactic acid), which has a fairly regular crystalline structure and can be made into fibres. Its properties can also be modified by mixing it in different proportions with the polymer poly-(D-lactic acid) formed from its optical isomer, D-lactic acid.
- PLA is used in biomedical applications (e.g. medical stitches), where its biodegradability can be utilised.

 Quick check 5

- It can also be incorporated into packaging, disposable tableware, etc. where its biodegradability means that it will not be a long-term litter problem.

QUICK CHECK QUESTIONS

1 Identify the repeat units and the linkages formed in the polymers (a) to (c) below.

(a)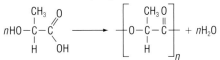
··OCH₂CH₂—O—C—CH₂CH₂—C—O—CH₂CH₂—O—C—CH₂CH₂—C—O—CH₂CH₂—O··

(b) ··C—(CH₂)₄—C—O—(CH₂)₆—O—C—(CH₂)₄—C—O—(CH₂)₆—O—C—(CH₂)₄—C··

(c) ··C—⬡—C—NH—⬡—NH—C—⬡—C—NH—⬡—NH—C—⬡—C··

2 For each polymer in question 1, identify the *two* monomer units present.

3 What are the problems associated with many polymers?

4 (a) Why is poly(lactic acid) a sustainable polymer?
  (b) State and explain some of its uses.

5 Glycolic acid ($HOCH_2COOH$) is a naturally occurring organic acid that can be extracted from corn starch. Show the structure of the repeating unit for poly(glycolic acid) and the equation for its formation.

Organic synthesis including chiral synthesis

Key words

- optical isomerism
- chirality
- stereoselectivity
- pharmaceuticals
- chiral synthesis
- chiral auxiliaries

✓ *Quick check 1*

✓ *Quick check 2*

Enzymes can be seen as **stereoselective** or **chiral** catalysts in that they will recognise or produce only one of the two optical isomers of a chiral compound.

When chiral molecules are involved, living systems use only one of the **optical isomers** of a chiral compound. Therefore **pharmaceutical** products acting on living systems often require the synthesis of these single optical isomers. As well as the effectiveness of the drug or chemical (e.g. perfume) other factors have to be taken into account. These are summarised in the table below:

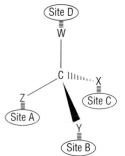

Sites A to D on the enzyme can recognise the spatial orientations of the groups W to Z on the molecule

Factor	Explanation	Example
Effectiveness	One isomer could exert a much stronger effect than the other could. Therefore, if it is used on its own it will be more pharmacologically effective in lower doses.	One optical isomer of the drug L-dopa is much more effective than the other. Administering the more effective isomer on its own is less wasteful and causes fewer side effects.
Toxicity	One isomer could have a beneficial effect whilst its mirror image could well be toxic.	The drug thalidomide is a chiral compound. One optical isomer stops morning sickness in pregnant women; the other causes deformities in unborn fetuses.

Chiral synthesis of single optical isomers

If a mixture of chiral compounds is administered as a drug to a patient, then possible toxic or undesirable side effects can leave the company responsible open to litigation. Therefore, it is important that only one optical isomer (the one required) is synthesised. To get round the problem of synthesising a mixture of optical isomers, chemists have invented different **chiral syntheses**, of which there are several types, as shown in the table below:

Method of chiral synthesis	How does it work?
Using transition metal complexes	In this method a transition metal complex acts as a stereoselective catalyst, giving a single optical isomer.
Chiral auxiliaries	A chiral auxiliary is a group that can be attached to a reactant molecule at some stage in the synthesis. This makes the synthesis follow one stereochemical route, producing just one optical isomer. At the end of the process the chiral auxiliary is removed.
Biosynthesis	Enzymes are chiral catalysts and are stereoselective, so will produce only one optical isomer of a chiral molecule.
Chiral pool synthesis	Using naturally chiral molecules such as amino acids and sugars in the synthetic process. The end product required must have a similar chirality to the naturally occurring chiral molecule.

Organic synthesis

The interrelationships between the different functional groups for both aliphatic and aromatic carbon compounds are summarised in the flowcharts opposite. These flowcharts are a useful summary of much of the carbon chemistry you have learned on the AS and A2 Chemistry courses.

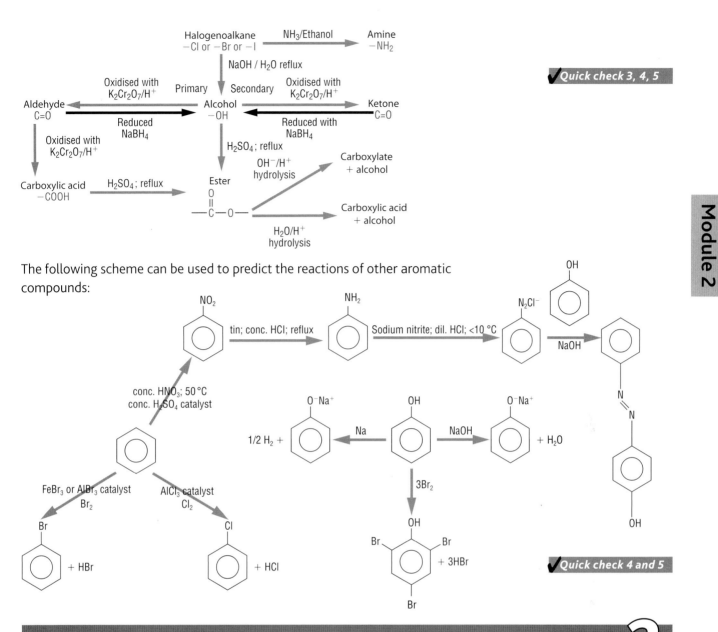

The following scheme can be used to predict the reactions of other aromatic compounds:

✓ *Quick check 3, 4, 5*

✓ *Quick check 4 and 5*

QUICK CHECK QUESTIONS

1 Why is it that natural systems are able to synthesise pure optical isomers of different compounds?

2 Give *two* reasons why it is important to synthesise the correct optical isomer when preparing a drug.

3 Explain how to carry out the following conversions.
   (a) $CH_3COOC_2H_5 \rightarrow CH_3CH_2OH \rightarrow CH_3COOH$
   (b) $CH_3CH_2CHO \rightarrow CH_3CH_2CH_2OH \rightarrow$
   $\qquad\qquad\qquad\qquad\qquad CH_3COOCH_2CH_2CH_3$

4 For the following conversions, identify the possible missing compounds.

   (a) $CH_3CHO \xrightarrow{\text{NaBH}_4 \text{ in water}} A \xrightarrow{\text{Reflux with acidified K}_2\text{Cr}_2\text{O}_7} B \xrightarrow[\text{reflux}]{\text{CH}_3\text{CH}_2\text{OH; H}^+ \text{ catalyst}} C$

   (b)

5 The structure of testosterone is shown below.
   (a) What are the functional groups present?
   (b) Which functional group will react with:
      (i) bromine in the dark?
      (ii) ethanoic acid?
      (iii) sodium borohydride in water?

27

Chromatography

Key words

- chromatography
- mobile phase
- stationary phase
- relative solubility
- adsorption
- solutes
- thin-layer chromatography
- gas chromatography
- GC–MS

General facts about chromatography

In this a **mixture** carried by a **mobile phase** is moved over a **stationary phase**.

The stationary phase tends to hold back the components of the mixture to differing extents, so they move through the stationary phase at different rates and therefore separate. The components of the mixture to be separated are called **solutes**.

Chromatography works by two mechanisms – **relative solubility** and **adsorption**.

✓ *Quick check 1*

	Mechanism	
	Relative solubility	**Adsorption**
Stationary phase	A solid or a viscous liquid adsorbed onto a solid	A polar solid that has to be dry
Mobile phase	A gas or liquid	A liquid
How does it work?	The solute (in the mixture) is more soluble in one phase than another. The components are separated according to their different solubilities in the two phases. If a solute is relatively more soluble in the mobile phase than another solute, then it will pass more quickly.	The components in the mixture are usually polar compounds and are attracted to the solid phase to varying degrees. The more the solute is attracted to the solid phase, the more slowly it will move.

✓ *Quick check 2*

The two chromatographic techniques covered are thin-layer chromatography (TLC) and gas chromatography (GC).

Thin-layer chromatography

Components	How it works/comments
Dry silica or alumina coated onto a glass or plastic surface is the stationary phase.	If the solid is absolutely dry, then the chromatography proceeds via adsorption. The solids are polar and therefore the speed at which the solutes move up the solid depends on their polarity.
The mobile phase can be a pure liquid or a mixture of liquids.	The chromatography depends on the relative solubilities of the solutes in the mobile and stationary phases.

TLC can be used to separate and identify non-volatile components.

R_f values in TLC

In TLC the movement of any solute compared to the solvent is called the R_f value. This value means that a solute can be identified even if the chromatogram is run for different times. As long as the substances used for the stationary and mobile phases are the same, a comparison is valid.

The components of a mixture can be identified by comparing their R_f values with those of known compounds, or by running the unknown mixture with the known compounds on the same chromatogram.

✓ *Quick check 2*

$$\text{The } R_f \text{ value} = \frac{distance\ moved\ by\ solute}{distance\ moved\ by\ solvent}$$

The distance moved by the solute is measured to the centre of the spot. For example, a solute moves 8.00 cm whilst the solvent moves 12.0 cm. The R_f value = 8/12 = 0.667.

Gas chromatography

Components	How it works/comments
The stationary phase is a solid or liquid adsorbed onto the surface of an inert solid on a long (1–30m) chromatography column.	Works by relative solubility. The polarity of the liquid on the column can be changed and hence changes the solubility of the solutes in the stationary phase. This can be used to achieve a better separation.
The mobile phase is a pure gas, like nitrogen, or a volatile liquid. The mobile phase is called the **carrier gas**.	The time taken for the volatile components of the mixture to move through the column is called their retention time.

Gas chromatograms

A gas chromatogram is a very sensitive way of separating and identifying the components of a mixture of volatile components. First, pure substances are run under identical conditions and their retention times measured. The mixture is then run and the retention times of the components compared with those of the pure substances.

Disadvantages of the method are shown below:
- A mixture may have many components and each of these has to have reference retention times for comparison.
- Unknown compounds may have no reference times for comparison and hence identification.
- Different substances may have identical retention times.
- Substances that have high boiling points cannot be separated.

✔ *Quick check 3*

✔ *Quick check 3*

Module 3

Analysis in GC

As each component of the mixture leaves the column, its relative concentration can be measured by a recorder. The relative amounts of each component in a mixture can be measured by the areas of their respective peaks.

% of each component = (area of peak/total area of peaks) × 100%

Linking the column to a **mass spectrometer** means that as the unknown leaves the column, it can be identified by comparing its fragmentation pattern with that of known compounds, and it can be identified almost instantly. **This technique is called gas chromatography–mass spectrometry (GC–MS)** and has the uses shown here:

Situation	Substances detected and identified
Environmental analysis	Organic pollutants in water; pesticides in food
Forensic drug detection	Minute amounts of drugs in forensic samples
Security operations	Explosives in luggage or on people
Space probes	The components of a planet's atmosphere

QUICK CHECK QUESTIONS

1 Describe the differences between chromatography based on adsorption and relative solubility.
2 The table below shows the results from TLC of three substances, P, Q and R, using the same solvent. Calculate their R_f values and then identify the unknown.

Substance	Distance moved /cm	Distance moved by solvent /cm
P	6.3	10
Q	5.6	7
R	6.5	9
Unknown	4.8	6

3 (a) The following retention times were obtained when a mixture was separated using GC: ethanol, 5 s; hexane, 10 s.
   (i) What do you think would happen if the chromatograph was run again but this time the silica on the column was coated with a more polar stationary phase?
   (ii) Explain your answer.
  (b) Give the advantages and disadvantages of gas–liquid chromatography.

Carbon-13 NMR

Key words

- NMR
- chemical shift
- isotope

A nuclear magnetic resonance (NMR) spectrum is obtained when the atomic nuclei of a material interact with the low-energy radiowave region of the electromagnetic spectrum in the presence of a very strong magnetic field.

Not every element has nuclei that have odd numbers of nucleons and residual magnetic spin resulting in an NMR spectrum. Two that do are ^1H and ^{13}C.

There are two main pieces of information that can be gleaned from a ^{13}C NMR spectrum (see table).

What it tells you	How this is shown on the spectrum
The number of chemically different carbon atoms present	The number of peaks
The chemical environment of the carbon	The value of the chemical shift (δ) in ppm. Each type of carbon has a characteristic range over which it can be located on the NMR spectrum.

✔ Quick check 1

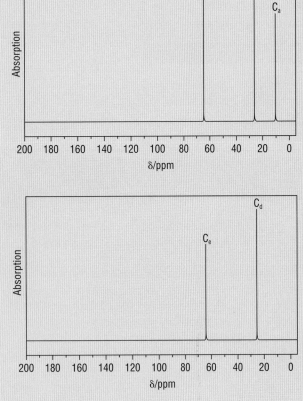

The isotope of carbon, ^{13}C, will produce NMR spectra at a different range of radiofrequencies from proton NMR (see page 32). Also the range of chemical shifts (0–200 ppm) is about 20 times larger than that for proton NMR spectra (0–10 ppm).

■ WORKED EXAMPLE

There are two alcohols with the molecular formula C_3H_8O. Their ^{13}C-NMR spectra are given on the left. Identify the alcohol responsible for each ^{13}C-NMR spectrum and explain the spectra.

STEP 1 Draw the two isomers and label the different types of carbon present. In this case the two isomers are propan-1-ol and propan-2-ol.

STEP 2 Analyse the spectrum to see how many peaks there are.

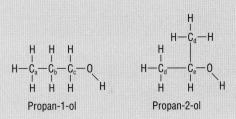

Propan-1-ol   Propan-2-ol

Spectrum A has three peaks and is therefore the spectrum for propan-1-ol because this has three types of carbon present relative to the –OH group. Spectrum B with its two peaks is that for propan-2-ol, with its two types of carbon.

The carbons responsible for each peak are shown on the diagram. When electronegative atoms are present, in general, the chemical shift increases with decreasing electron density around the carbon.

Typical carbon-13 chemical shifts

The figure shows typical chemical shifts for carbon-13 in different types of chemical environment. These can change with the solvent used, the concentration and what other chemical groups are present.

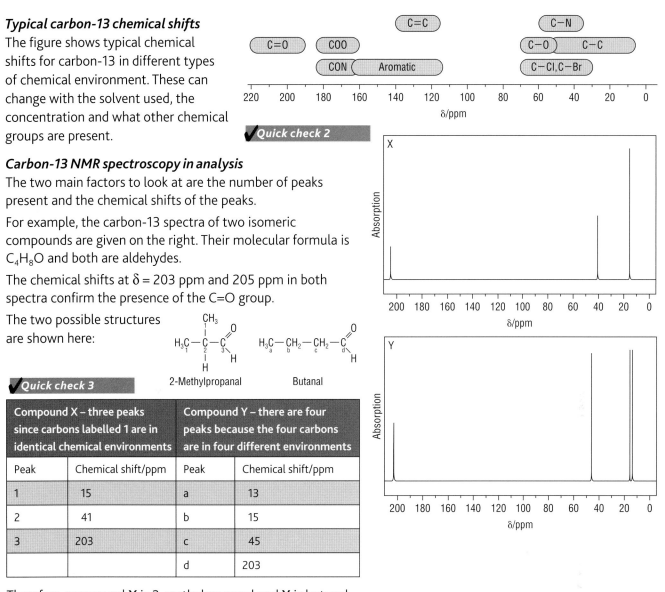

✔Quick check 2

Carbon-13 NMR spectroscopy in analysis

The two main factors to look at are the number of peaks present and the chemical shifts of the peaks.

For example, the carbon-13 spectra of two isomeric compounds are given on the right. Their molecular formula is C_4H_8O and both are aldehydes.

The chemical shifts at $\delta = 203$ ppm and 205 ppm in both spectra confirm the presence of the C=O group.

The two possible structures are shown here:

2-Methylpropanal   Butanal

✔Quick check 3

Compound X – three peaks since carbons labelled 1 are in identical chemical environments		Compound Y – there are four peaks because the four carbons are in four different environments	
Peak	Chemical shift/ppm	Peak	Chemical shift/ppm
1	15	a	13
2	41	b	15
3	203	c	45
		d	203

Therefore compound X is 2-methylpropanal and Y is butanal.

The further away the carbon is from the electronegative oxygen, the smaller the chemical shift.

QUICK CHECK QUESTIONS

1 Give the number of peaks obtained and the characteristic chemical shifts (in ppm) for the ^{13}C NMR spectra of the following compounds.

(a)   (b)   (c)   (d)

$H_3C-\overset{O}{\overset{||}{C}}-CH_3$ $H_3C-C\overset{\nearrow O}{\underset{\searrow OH}{}}$ $H_2C=CH_2$

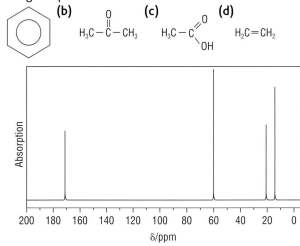

2 The two ^{13}C NMR spectra (below) are for the compounds $CH_3CH(CH_3)COOH$ and $CH_3COOCH_2CH_3$. Identify which spectrum is for which compound and explain your choice.

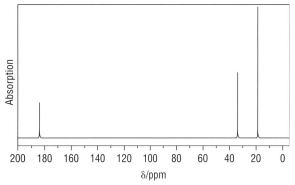

Module 3

Proton NMR

Key words

- magnetic resonance imaging
- chemical shift
- splitting
- tetramethylsilane

Examiner tip

An example of a **low-resolution** (no splitting shown) spectrum is that of ethanol, **CH₃CH₂OH** (see diagram). This spectrum has three peaks because of the three types of proton present and their positions in the spectrum are indicative of their position in the molecule relative to the oxygen of the –OH group. The areas under the three peaks also show the relative numbers of each type of proton, i.e. 3 for the CH₃ protons, 2 for the CH₂ protons and 1 for the OH proton.

Understanding the concept

You *do not* have to know why or how NMR spectra come about for your exam, but it would aid your understanding and interpretation of this type of spectroscopy if you knew a few background facts. An NMR spectrum is obtained when a material interacts with the low-energy radiowave region of the electromagnetic spectrum in the presence of a very strong magnetic field.

NMR technology is the same as that used in **magnetic resonance imaging (MRI)** to obtain an image of soft tissue inside the body. This technique is very valuable because it is non-invasive and the low-energy radiowaves used do not damage body tissues. Different biochemical tissues (e.g. bone and fat) have different numbers of protons and hence give signals of different intensities.

There are four main features of a proton NMR spectrum:

1 the position of the absorbance peaks (the δ value or **chemical shift**)
2 the number of peaks
3 the relative peak area, shown by an integration value
4 the **splitting** of each absorbance peak into smaller, finer peaks. For example, a peak split into two smaller peaks is called a doublet, one split into three peaks a triplet, etc.

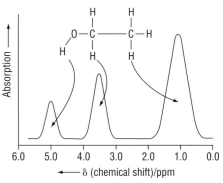

Features 3 and 4 are not present in C-13 NMR spectra.

NMR is a very powerful technique. The table below summarises what it can tell you about a compound and how this is shown in its NMR spectrum.

What it tells you	How this is shown in the spectrum
How many types of proton present	The number of distinct peaks = number of types of proton
How many of each type of proton	The relative area under each peak = relative number of protons
What type of proton	The value of δ and then use of tables. See below.
The number of chemically different protons on the atom adjacent to a particular type of proton e.g. –CH₂CH₃	By the splitting of the distinct peaks. If there are *n* chemically different protons on the carbon (or other atom) adjacent to a particular type of proton, then the peak for that proton is split into *n* + 1 peaks. For example, the peak for the methyl (CH₃) protons in –CH₂CH₃ will be split into a triplet by the **two** chemically different protons on the adjacent –CH₂– group.

✓*Quick check 1*

Examiner tip

TMS is the abbreviation for tetramethylsilane, the most common internal standard used in NMR spectra.

How can we relate NMR spectra to structures?

Note: every proton NMR spectrum will have a peak at δ = 0 ppm due to the presence of protons from tetramethylsilane (Si(CH₃)₄), the internal standard used.

■ WORKED EXAMPLE

Here are the spectra of two compounds having the molecular formula C_2H_6O.

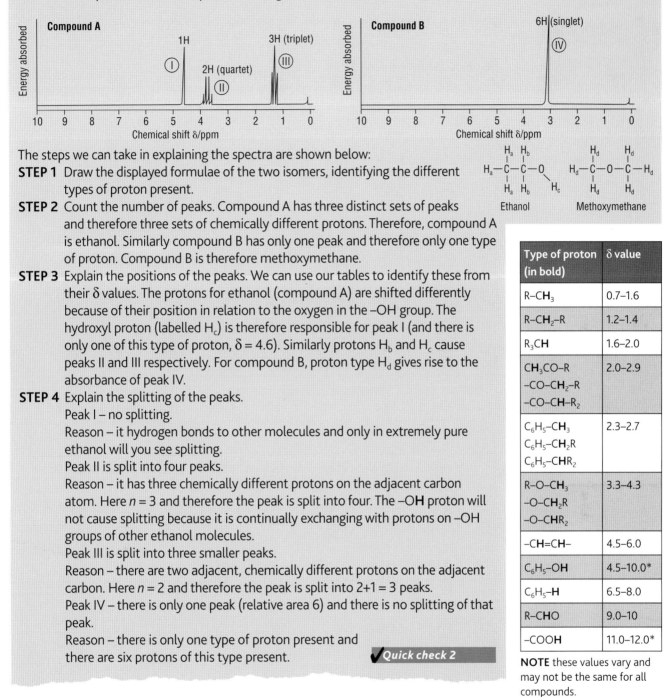

The steps we can take in explaining the spectra are shown below:

STEP 1 Draw the displayed formulae of the two isomers, identifying the different types of proton present.

STEP 2 Count the number of peaks. Compound A has three distinct sets of peaks and therefore three sets of chemically different protons. Therefore, compound A is ethanol. Similarly compound B has only one peak and therefore only one type of proton. Compound B is therefore methoxymethane.

STEP 3 Explain the positions of the peaks. We can use our tables to identify these from their δ values. The protons for ethanol (compound A) are shifted differently because of their position in relation to the oxygen in the –OH group. The hydroxyl proton (labelled H_c) is therefore responsible for peak I (and there is only one of this type of proton, $\delta = 4.6$). Similarly protons H_b and H_c cause peaks II and III respectively. For compound B, proton type H_d gives rise to the absorbance of peak IV.

STEP 4 Explain the splitting of the peaks.

Peak I – no splitting.

Reason – it hydrogen bonds to other molecules and only in extremely pure ethanol will you see splitting.

Peak II is split into four peaks.

Reason – it has three chemically different protons on the adjacent carbon atom. Here $n = 3$ and therefore the peak is split into four. The –**OH** proton will not cause splitting because it is continually exchanging with protons on –OH groups of other ethanol molecules.

Peak III is split into three smaller peaks.

Reason – there are two adjacent, chemically different protons on the adjacent carbon. Here $n = 2$ and therefore the peak is split into 2+1 = 3 peaks.

Peak IV – there is only one peak (relative area 6) and there is no splitting of that peak.

Reason – there is only one type of proton present and there are six protons of this type present.

✔ *Quick check 2*

Ethanol — Methoxymethane

Type of proton (in bold)	δ value
R–CH_3	0.7–1.6
R–CH_2–R	1.2–1.4
R_3CH	1.6–2.0
CH_3CO–R –CO–CH_2–R –CO–CH–R_2	2.0–2.9
C_6H_5–CH_3 C_6H_5–CH_2R C_6H_5–CHR_2	2.3–2.7
R–O–CH_3 –O–CH_2R –O–CHR_2	3.3–4.3
–CH=CH–	4.5–6.0
C_6H_5–**OH**	4.5–10.0*
C_6H_5–**H**	6.5–8.0
R–**CHO**	9.0–10
–**COOH**	11.0–12.0*

NOTE these values vary and may not be the same for all compounds.

* These values are sensitive to solvent, substituents and concentration.

QUICK CHECK QUESTIONS

1 What information do the following features of an NMR spectrum tell you?
   (a) There are four distinct peaks.
   (b) One peak is split into a triplet.

2 The NMR spectrum shown is that for one of the isomers of the compound $C_2H_4Cl_2$. The number above the peak reflects its relative area.
   (a) Draw displayed formulae for the two possible isomers of the compound.
   (b) Name the isomer responsible for the spectrum.
   (c) Explain your answer to part (b).

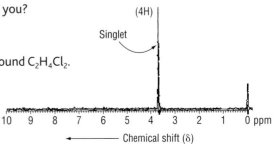

UNIT 1

Identifying compounds using NMR spectroscopy

Key words

- labile (exchangeable) protons

This section will show how powerful this method is in determining chemical structures.

■ WORKED EXAMPLE

The NMR spectrum for an isomer of the carbonyl compound C_4H_8O is given below. Identify the isomer and explain your reasoning.

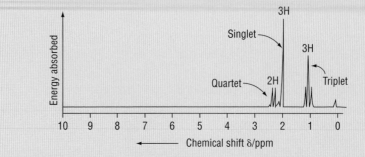

STEP 1 Draw the displayed formulae for all the possible structural isomers of the compound.

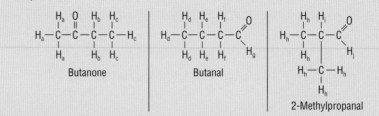

Butanone Butanal 2-Methylpropanal

STEP 2 Identify the compound.
The three peaks in the spectrum show that there are *three chemically different types of proton*.
Butanone has three types (H_a, H_b and H_c) and 2-methylpropanal also has three (types H_h, H_i and H_j). Butanal, however, has four types (H_d, H_e, H_f and H_g). This narrows the choice to butanone and 2-methylpropanal. However, none of the peaks indicate the presence of a proton corresponding to an aldehyde proton (chemical shift $\delta = 9.0$–10 ppm) and therefore the compound must be *butanone*.

STEP 3 Explain the spectrum.
The singlet at $\delta = 2.1$ ppm. There are three protons (H_a) and there are no chemically different protons on the adjacent carbon, hence no splitting ($n = 0$, and $n + 1 = 1$; hence only one peak is produced).
The quartet at $\delta = 2.5$ ppm. This is produced by the two (relative peak area = 2) H_b protons and the splitting into a quartet is caused by the three chemically different protons on the adjacent carbon (the three H_c protons).
The triplet at $\delta = 1.0$ ppm. There are three such protons and the peak is split into a triplet by the two adjacent H_b protons.

✓ *Quick check 1, 2, 3*

The δ values of the protons reflect their chemical environment.

Exchangeable (labile) protons

Protons attached to –OH, –COOH and –NH$_2$ groups can interchange with the protons in water. They are called **labile or exchangeable protons**.

- Adding the compound in question to deuterium oxide, D$_2$O, can identify such protons. For example, with an alcohol, ROH, the following exchange occurs.

$$\text{ROH} \quad + \quad \text{D}_2\text{O} \quad \rightleftharpoons \quad \text{ROD} \quad + \quad \text{HOD}$$
Peak present at δ =4.5   No peak at δ = 4.5

- The proton responsible for one of the peaks in the NMR spectrum of the alcohol has been removed and the peak itself disappears.

- This will not happen with protons that do not exchange with water, such as methyl, –CH$_3$, protons.

The use of CDCl$_3$ as a solvent

Many solvents required for dissolving samples contain protons. The signals produced by these protons would interfere with and confuse the interpretation of the spectra. To overcome this problem, solvents such as CDCl$_3$ (not CHCl$_3$) are used, which are still good solvents but, because they contain deuterium instead of ^{1}H, will not produce an NMR signal.

Module 3

QUICK CHECK QUESTIONS

Note: For the spectra given below, the peak at δ = 0 ppm is due to the absorbance of the standard protons on TMS.

1 There are two possible carbonyl isomers of the compound C$_3$H$_6$O.
   (a) Draw displayed formulae for these two isomers.
   (b) The NMR spectrum for one of these isomers has three peaks. One of the peaks, at δ = 9.7 ppm, is split into a triplet. Name the isomer responsible for the spectrum and explain your answer.

2 Predict the main features of the proton NMR spectrum of ethanal. Your answer should include reference to the number of peaks; the relative area under each peak; the chemical shift of each peak; and the splitting patterns obtained.

3 The proton NMR spectrum below is that of ethanoic acid.
   (a) Draw the displayed formula of the compound.
   (b) Explain why there are two peaks.
   (c) Which peak is for which type of proton?
   (d) Which peak will disappear if the acid is mixed with D$_2$O?

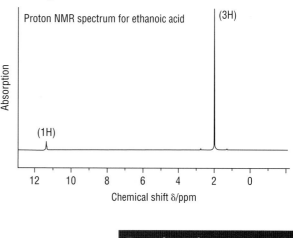

Combined analytical techniques

Combined techniques in identifying carbon compounds

In some exam questions and in investigations you have to identify a carbon compound from chemical tests and by using the three sets of spectra available. The procedure you can use is shown below:

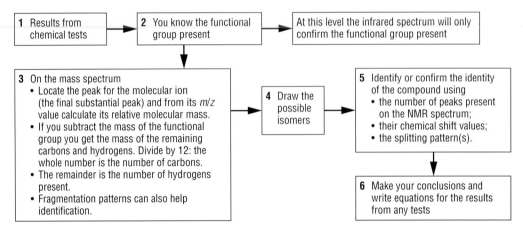

| 1 Results from chemical tests | → | 2 You know the functional group present | → | At this level the infrared spectrum will only confirm the functional group present |

3 On the mass spectrum
- Locate the peak for the molecular ion (the final substantial peak) and from its *m/z* value calculate its relative molecular mass.
- If you subtract the mass of the functional group you get the mass of the remaining carbons and hydrogens. Divide by 12: the whole number is the number of carbons.
- The remainder is the number of hydrogens present.
- Fragmentation patterns can also help identification.

4 Draw the possible isomers

5 Identify or confirm the identity of the compound using
- the number of peaks present on the NMR spectrum;
- their chemical shift values;
- the splitting pattern(s).

6 Make your conclusions and write equations for the results from any tests

✓ *Quick check 1*

An example of using combined techniques to identify an unknown compound:

An unknown compound X gave a yellow-orange precipitate with 2,4-dinitrophenylhydrazine and when tested with Tollens' reagent it gave a silver mirror.

Using this information we can see that:

X is a carbonyl compound containing the >C=O group. Therefore it is an *aldehyde* or a *ketone*. This is shown by the reaction with 2,4-dinitrophenylhydrazine.

X is an *aldehyde* with the –CHO group. The silver mirror with Tollens' reagent is given by aldehydes but not by ketones.

The infrared, mass and proton NMR spectra of X are given below and opposite:

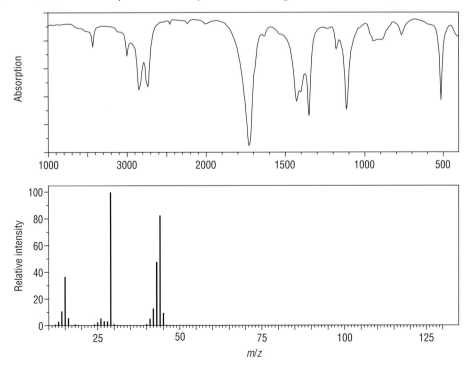

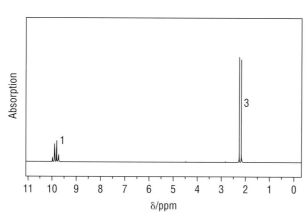

The infrared spectrum shows a strong absorption at 1730 cm^{-1}, confirming the presence of the carbonyl (>C=O) group.

✓*Quick check 2*

In the mass spectrum the molecular ion peak is at $m/z = 44$.

Therefore the relative molecular mass is 44.

If we subtract the mass of the aldehyde (CHO) group from this, we are left with 44 – 29, which is equal to 15.

To calculate the number of carbons in the remainder of the molecule we divide by 12, giving us one carbon, and the remaining three mass units are due to three hydrogens. There are no isomers.

The compound is therefore ethanal – CH$_3$CHO.

The NMR spectrum confirms this because there are *two* peaks and the chemical shift of one ($\delta = 9.8$ ppm) is for an aldehyde C**H**O proton. This peak is a quartet, because of the *three* chemically different protons (–C**H**$_3$) on the adjacent carbon. The other peak is a doublet because of the *one* chemically different proton on the adjacent –C**H**O group.

QUICK CHECK QUESTIONS

1   For the mass spectrum of ethanal, explain why there are peaks at $m/z = 15$ and 29. Give equations to support your answer.

2   The displayed formulae of the three isomeric ketones with the molecular formula C$_5$H$_{10}$O are shown below.

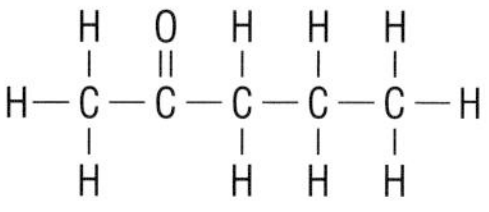

```
    H  O  H  H  H
    |  ||  |  |  |
H — C — C — C — C — C — H
    |     |  |  |
    H     H  H  H
          A
```

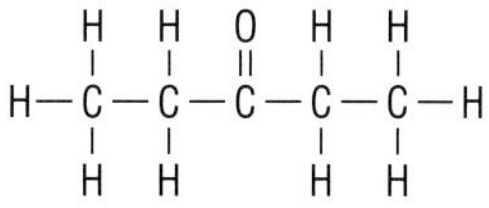

```
    H  H  O  H  H
    |  |  ||  |  |
H — C — C — C — C — C — H
    |  |     |  |
    H  H     H  H
          B
```

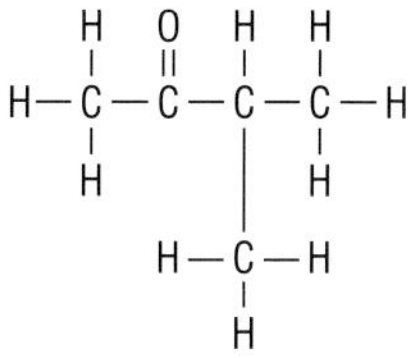

```
    H  O  H  H
    |  ||  |  |
H — C — C — C — C — H
    |     |  |
    H     |  H
       H — C — H
           |
           H
          C
```

(a) What absorption will all three show on their infrared spectra?

(b) Explain how they can be distinguished using their ^{13}C-NMR spectra.

(c) A compound Y gave mass and NMR spectra with the characteristics below. Which of the three ketones is Y?

Spectrum	Characteristics
Mass spectrum	A molecular ion peak at $m/z = 86$; a fragment with an m/z value of 43
Proton-NMR	There are three sets of peaks: a singlet at $\delta = 2.2$ ppm (three protons); a septet (seven peaks) at $\delta = 2.6$ ppm, and a doublet at $\delta = 1.3$ ppm (six protons).
^{13}C-NMR	There are four peaks .

1 A hydrocarbon is known to contain a benzene ring. It has a relative molecular mass of 106 and has the following composition by mass: C, 90.56%; H, 9.44%.
   (a) (i)   Use the data above to show that the empirical formula is C_4H_5.
       (ii) Deduce the molecular formula.
       (iii) Draw structures for all possible isomers of this hydrocarbon that contain a benzene ring.   [7]
   (b) In the presence of a catalyst (such as aluminium chloride), one of these isomers (A) reacts with chlorine to give only *one* monochloro-product, B.
       (i)   Deduce which of the isomers in (a)(iii) is **A**.
       (ii) Draw the structure of B.   [2]
                                                                   [TOTAL 9 marks]

2 The diagram below shows the three amino acids, glycine, alanine and phenylalanine.

$$H_2N-CH_2-\underset{\underset{OH}{|}}{\overset{\overset{O}{\|}}{C}} \qquad H_2N-\underset{\underset{H}{|}}{\overset{\overset{CH_3}{|}}{C}}-\underset{\underset{OH}{|}}{\overset{\overset{O}{\|}}{C}} \qquad H_2N-\underset{\underset{H}{|}}{\overset{\overset{CH_2}{|}}{C}}-\underset{\underset{OH}{|}}{\overset{\overset{O}{\|}}{C}}$$

   (a) Explain why glycine does *not* have optical isomers but the other two amino acids do.   [1]
   (b) Amino acids can react both with acids and with bases, and are also capable of forming zwitterions.
       Write equations for the reactions between
       (i)   glycine and HCl
       (ii) alanine and NaOH.   [2]
   (c) The pH at which an amino acid forms its zwitterion is different for each amino acid and is known as the isoelectric point.
       (i)   The isoelectric point for glycine is at pH = 5.97. Draw the zwitterion formed by glycine at this pH.   [1]
       (ii) The isoelectric points of alanine and phenylalanine are at pH 6.00 and at pH 5.48, respectively. Draw the displayed formulae of the predominant forms of the amino acids present at pH 5.7.   [2]
   (d) Glycine and alanine can react together to form *two* dipeptides.
       Draw the displayed formula of both dipeptides that could be formed from the reaction between glycine and alanine.   [4]
                                                                   [TOTAL 10 marks]

3 There are four aldehydes and three ketones with the molecular formula $C_5H_{10}O$.
   (a) Draw the structural formulae of all seven of these isomers.   [7]
   (b) The carbon-13 and proton NMR spectra of one of these isomers are shown opposite.
       Using these spectra, identify the isomer to which they belong and explain your choice.   [5]
   (c) Explain how you could identify this isomer chemically, using Tollens' reagent and 2,4-dinitrophenylhydrazine.   [7]
   (d) Give the mechanism for the reduction of this isomer with sodium borohydride in the presence of water.   [4]
                                                                   [TOTAL 23 marks]

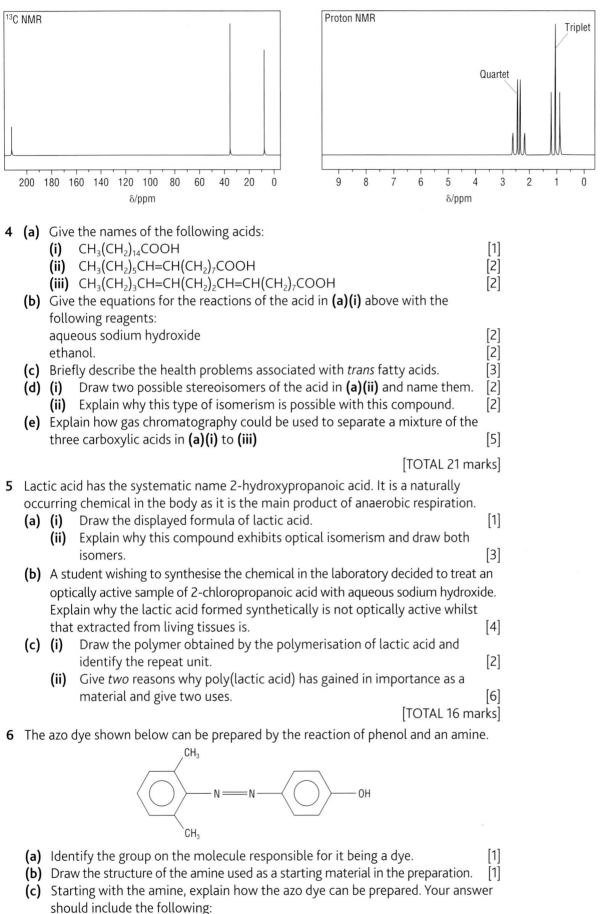

4 **(a)** Give the names of the following acids:
   (i) $CH_3(CH_2)_{14}COOH$ [1]
   (ii) $CH_3(CH_2)_5CH=CH(CH_2)_7COOH$ [2]
   (iii) $CH_3(CH_2)_3CH=CH(CH_2)_2CH=CH(CH_2)_7COOH$ [2]
   (b) Give the equations for the reactions of the acid in **(a)(i)** above with the following reagents:
   aqueous sodium hydroxide [2]
   ethanol. [2]
   (c) Briefly describe the health problems associated with *trans* fatty acids. [3]
   (d) (i) Draw two possible stereoisomers of the acid in **(a)(ii)** and name them. [2]
   (ii) Explain why this type of isomerism is possible with this compound. [2]
   (e) Explain how gas chromatography could be used to separate a mixture of the three carboxylic acids in **(a)(i)** to **(iii)** [5]

   [TOTAL 21 marks]

5 Lactic acid has the systematic name 2-hydroxypropanoic acid. It is a naturally occurring chemical in the body as it is the main product of anaerobic respiration.
   (a) (i) Draw the displayed formula of lactic acid. [1]
   (ii) Explain why this compound exhibits optical isomerism and draw both isomers. [3]
   (b) A student wishing to synthesise the chemical in the laboratory decided to treat an optically active sample of 2-chloropropanoic acid with aqueous sodium hydroxide. Explain why the lactic acid formed synthetically is not optically active whilst that extracted from living tissues is. [4]
   (c) (i) Draw the polymer obtained by the polymerisation of lactic acid and identify the repeat unit. [2]
   (ii) Give *two* reasons why poly(lactic acid) has gained in importance as a material and give two uses. [6]

   [TOTAL 16 marks]

6 The azo dye shown below can be prepared by the reaction of phenol and an amine.

   (a) Identify the group on the molecule responsible for it being a dye. [1]
   (b) Draw the structure of the amine used as a starting material in the preparation. [1]
   (c) Starting with the amine, explain how the azo dye can be prepared. Your answer should include the following:
   the sequence of reactions, including equations
   intermediate reagents used and conditions. [9]

   [TOTAL 11 marks]

Equilibria, energetics and elements

Module 1 – Rates, equilibrium and pH, pages 42–63

You will need to have revised the work you did on rates, equilibria and acids to be comfortable when you tackle this module. This module is more quantitative (there is a lot more number work involved) than the one you did last year, so you must be familiar with your calculator and its mode of working. If you do not do A-level mathematics then ask someone about logarithms (logs) and how to use them.

Double-page spread	Specification	Previous knowledge that you will use in this module
How fast? The rate equation	5.1.1	Work on rates, gradients of graphs as a means of measuring the rate; methods of monitoring how quickly a reaction is proceeding
How fast? Calculating k, the effect of T and reaction mechanisms	5.1.1	Make sure you can rearrange formulae and know about how to calculate units. Remember what is meant by a mechanism from your carbon chemistry modules
How fast? Concentration–time graphs	5.1.1	Lines and curves from GCSE mathematics
How fast? Rate–concentration graphs	5.1.1	Lines and curves from GCSE mathematics
How far? The equilibrium law and K_c	5.1.2	Revise equilibria and Le Chatelier's principle
How far? How to calculate the value of K_c	5.1.2	Revise equilibria and Le Chatelier's principle
Acids and bases	5.1.3	The reactions of acids
Acids and bases in aqueous solution	5.1.3	Remember what the pH scale measures
The chemistry of weak acids	5.1.3	Review your work on equilibria and K_c
Buffers – how they work and their pH	5.1.3	Use your knowledge of pH and equilibria
Indicators, acid–base titration curves and neutralisation	5.1.3	Calculations for titrations and moles in solutions

Module 2 – Energy, pages 64–77

Double-page spread	Specification	Previous knowledge that you will use in this module
Lattice enthalpy	5.2.1	Enthalpy and ionic bonding
The Born–Haber cycle	5.2.1	Hess cycles
Things to know about Born–Haber cycles and hydration enthalpies	5.2.1	Hess cycles, electrostatic attractions between polar molecules and ions
Enthalpy and entropy	5.2.2	Hess cycles and energetics
Redox reactions and electrode potentials	5.2.3	The reactivity series and redox reactions
Standard electrode potentials	5.2.3	Redox reactions
Fuel cells, fuel cell vehicles and the hydrogen economy	5.2.3	Here you will apply much of the above plus your work at AS-level on global warming

Module 3 – Transition elements, pages 78–87

Double-page spread	Specification	Previous knowledge that you will use in this module
Transition elements – electron configurations	5.3.2	Electron configurations from the first module in AS
Transition elements – oxidation states, catalytic behaviour and the hydroxides	5.3.2	Redox reactions, catalysts and ionic equations
Complex ions	5.3.2	Dative covalent bonding
Ligand substitution	5.3.2	K_c and equilibria
Redox reactions and titration calculations	5.3.2	Redox reactions, balancing equations and titrations based on moles in solutions

End-of-unit questions, pages 88–90

How fast? The rate equation

Key words

- rate equation
- rate constant
- order
- total (overall) order

Examiner tip

Units of concentration = mol dm^{-3}

Units of rate = mol dm^{-3} s^{-1} or mol dm^{-3} min^{-1} etc.

Examiner tip

Note that [A] and [B] are the concentrations at the *beginning* of the reaction. This means we are measuring the *initial* rate. Of course the rate of a reaction can change with time, getting slower towards the end of the reaction. Measuring the initial rate, the instant the reactants are mixed, means there is zero concentration of products that might cause a reverse reaction.

✔ *Quick check 1*

This part of the unit deals with *rates of reaction*.

From previous work, you know that

$$\text{rate of reaction} = \frac{\text{change in concentration}}{\text{time}}$$

The **rate equation** is a way of calculating the rate of a chemical reaction.

The rate equation has a general form for the reaction

A + B → products

rate = $k[\text{A}]^m[\text{B}]^n$

where:

- A and B are the *reactants*.
- [A] is the *initial* concentration of A.
- [B] is the *initial* concentration of B.
- k is a constant called the **rate constant.** k relates the rate of a chemical reaction with the reactant concentrations. *The larger the value of* k, *the faster the reaction goes.*
- m is the **order** of the reaction with respect to reactant A.
- n is the **order** of the reaction with respect to reactant B.

The order of a reactant is the power to which its concentration is raised in the rate equation.

The **total (overall) order** of the reaction is $m + n$, the sum of the orders for each reactant.

Each reaction has its own rate equation. *This can only be worked out experimentally.*

Examples:

$SO_2Cl_2(g) \rightarrow SO_2(g) + Cl_2(g)$	$BrO_3^-(aq) + 5Br^-(aq) + 6H^+(aq) \rightarrow 3Br_2(aq) + 3H_2O(l)$
rate = $k[SO_2Cl_2(g)]$	*rate* = $k[BrO_3^-][Br^-][H^+]^2$
The rate is:	The rate is:
• first order with respect to SO_2Cl_2	• first order with respect to BrO_3^-
• first order overall	• first order with respect to Br^-
	• second order with respect to H^+
	• fourth order overall

Units of the rate constant

Because each reaction has its own rate equation, the units for k change according to the rate equation.

You can work out the units of k.

STEP 1 Rearrange the rate equation to get k:

$$k = \frac{rate}{[\text{A}]^m[\text{B}]^n}$$

STEP 2 Substitute the units into this equation:

Units for rate are mol dm^{-3} s^{-1}.

Units for [A] and [B] are mol dm^{-3}.

STEP 3 Cancel any units that are the same on the top and bottom of the expression.

■ WORKED EXAMPLE 1

For the reaction

$$(CH_3)_3CBr(l) + H_2O(l) \rightarrow (CH_3)_3COH(l) + HBr(aq)$$

The rate equation is $rate = k[(CH_3)_3CBr]$.

This is a first-order rate equation. What are the units of k?

$$k = \frac{rate}{[(CH_3)_3\,CBr]} = \frac{mol\ dm^{-3}\ s^{-1}}{mol\ dm^{-3}} = s^{-1}$$

■ WORKED EXAMPLE 2

For the reaction $2ICl(g) + H_2(g) \rightarrow 2HCl(g) + I_2(g)$ the rate equation is $rate = k[ICl][H_2]$.

This is a second-order rate equation. What are the units of k?

$$k = \frac{rate}{[ICl][H_2]} = \frac{mol\ dm^{-3}\ s^{-1}}{mol\ dm^{-3} \times mol\ dm^{-3}}$$

$$= \frac{s^{-1}}{mol\ dm^{-3}},\ \text{which is written as } dm^3\ mol^{-1}\ s^{-1}$$

Examiner tip

You must write the units as $dm^3\ mol^{-1}\ s^{-1}$.

Do not leave them as $s^{-1}\ mol\ dm^{-3}$.

You can see that *the units of* k *depend on the order of the reaction*. You can work these out, or remember:

Order	Rate equation	Units of k
Zero order	$rate = k[A]^0$	$mol\ dm^{-3}\ s^{-1}$
First order	$rate = k[A]^1$	s^{-1}
Second order	$rate = k[A]^2$	$dm^3\ mol^{-1}\ s^{-1}$
Third order	$rate = k[A]^3$	$dm^6\ mol^{-2}\ s^{-1}$

 ✔*Quick check 2 and 3*

QUICK CHECK QUESTIONS

1 For the rate equation $rate = k[NO][O_3]$:
   (a) What is the order with respect to NO?
   (b) What is the total order of the reaction?

2 What are the units of k for the reaction in question 1?

3 The table below shows how the rate of reaction depends on various concentrations of reactants that we will simply call A, B and C. Draw out the table and fill in the gaps labelled (a) to (n).

Order with respect to [A]	Order with respect to [B]	Order with respect to [C]	Rate equation	Overall order	Units of k
1st	2nd	Zero	(a)	(b)	(c)
(d)	(e)	(f)	$rate = k[A][B]$	(g)	(h)
1st	2nd	Zero	(i)	(j)	(k)
Zero	Zero	2nd	(l)	(m)	(n)

43

UNIT 2

How fast? Calculating k, the effect of T and reaction mechanisms

Key words

- mechanisms
- rate-determining step

You must be able to calculate the rate of a reaction from the rate equation, given the value of k and initial concentrations. This simply means substituting into the rate equation.

As well as this, make sure you can calculate k, the rate constant, using the rate equation. This means you have to rearrange the rate equation to give:

$$k = \frac{rate}{[A]^m[B]^n}$$

■ WORKED EXAMPLE

For the reaction

$$2KI(aq) + K_2S_2O_8(aq) \rightarrow 2K_2SO_4(aq) + I_2(aq)$$

$$rate = k[KI]\,[K_2S_2O_8].$$

The initial concentration of KI is 1.00×10^{-2} mol dm^{-3}, and the initial concentration of $K_2S_2O_8$ is 5.00×10^{-4} mol dm^{-3}.

The initial rate of disappearance of $K_2S_2O_8$ is 1.02×10^{-8} mol dm^{-3} s^{-1}.

Calculate the rate constant.

STEP 1 Work out the equation to calculate k by rearranging the rate equation:

$$rate = k[KI]\,[K_2S_2O_8]$$

$$\text{so } k = \frac{rate}{[KI]\,[K_2S_2O_8]}$$

STEP 2 List the values you have been given for the rate, the concentration of KI and the concentration of $K_2S_2O_8$. Make sure the units are all correct.

$$rate = 1.02 \times 10^{-8} \text{ mol dm}^{-3} \text{ s}^{-1}$$

$$[KI] = 1.0 \times 10^{-2} \text{ mol dm}^{-3}$$

$$[K_2S_2O_8] = 5.0 \times 10^{-4} \text{ mol dm}^{-3}$$

STEP 3 Substitute into equation for k:

$$k = \frac{1.02 \times 10^{-8}}{(1.00 \times 10^{-2})(5.00 \times 10^{-4})}$$

STEP 4 Calculate k:

$$k = 2.04 \times 10^{-3}$$

STEP 5 Work out the units for k:

units are $\dfrac{\text{mol dm}^{-3} \text{ s}^{-1}}{(\text{mol dm}^{-3})(\text{mol dm}^{-3})}$, which is dm^3 mol^{-1} s^{-1}

Answer $k = 2.04 \times 10^{-3}$ dm^3 mol^{-1} s^{-1} to three sig. figs.

Examiner tip

You can see that this type of question involves lots of information. Don't be put off – just sort it out carefully.

Examiner tip

large k = fast reaction, small k = slow reaction

✔ *Quick check 1*

The effect of a temperature change on *k*

When the temperature of a reaction is increased the molecules move faster, so there are more collisions and more chance of a product being formed.

- A rise in temperature *increases* the value of *k*. This means the reaction goes faster.
- A fall in temperature *decreases* the value of *k*. This means the reaction slows down.

Reaction mechanisms

One of these steps may occur much more slowly than the others. This step is called the **rate-determining step (rds)**, because it determines how fast the reaction goes. It is the slowest step in a sequence of steps.

How do we know what the rate-determining step is? Well, that particular reaction is shown by the rate law.

Look at this example to understand:

$H_2(g) + 2ICl(g) \rightarrow I_2(g) + 2HCl(g)$

rate = [H_2] [ICl]

- The *stoichiometric equation* tells us that, *overall*, one molecule of H_2 reacts with two molecules of ICl to give the products. But it does not tell us about the individual steps which make up the overall reaction. *There is no link between the stoichiometric equation and the reaction mechanism or the rate equation*.
- The rate equation tells us that the rate-determining step involves one molecule of H_2 and one molecule of ICl, because it is first order with respect to both.

In this case the rds is the first step, and an intermediate, HI, is formed before the second, fast step gives the products. Don't worry if you can't work out what the intermediate is, as it can be rather obscure. You will only be asked very obvious rate-determining steps. Just concentrate on knowing how many molecules of each intermediate are involved in the rds.

Examiner tip

A reaction mechanism is a description of a chemical reaction as a series of one-step processes.

✓ *Quick check 2*

Examiner tip

In the rate law the order of reaction with respect to a reactant indicates how many molecules (or ions) of that reactant participate in the rate-determining step.

If a reactant is first order, one molecule reacts in the rds.

If a reactant is second order, two molecules react in the rds etc.

Examiner tip

The actual mechanism is:
$H_2 + ICl \rightarrow HI + HCl$
slow step

then
$HI + ICl \rightarrow I_2 + HCl$
fast step

Module 1

QUICK CHECK QUESTIONS

1 The rate of dissociation of the gas SO_2Cl_2 follows this rate law: *rate* = $k[SO_2Cl_2]$. If the initial rate is 1.35×10^{-4} mol dm^{-3} s^{-1} and the initial concentration of SO_2Cl_2 is 0.450 mol dm^{-3}, calculate the value of the rate constant *k*.

2 Suggest what the rate-determining step is in this reaction:

$CH_3COCH_3(aq) + I_2(aq) \rightarrow CH_3COCH_2I(aq) + HI(aq)$

rate = $k[CH_3COCH_3] [H^+] [I_2]^0$

How fast? Concentration–time graphs

When a reaction is taking place, it is possible to measure the concentration of a reactant over time. The *shape* of this *concentration–time graph* tells you the **order** with respect to the reactant being measured.

The following graphs show you the shape which each different order for a reactant gives. You must know these graphs – you can tell the order of a reactant just by looking at them. They all assume that there is one reactant only, A.

Zero order – the gradient of the graph never changes, showing that the rate is unaffected by the concentration of the reactant. The rate equation is: $rate = k[A]^0$	First order – the gradient of the graph changes, showing the concentration does affect the rate. Also, the half-life (the time taken for the concentration to halve) is a constant. The rate equation is: $rate = k[A]^1$	Second order – the concentration affects the rate but half-life is not constant and increases with time. The rate equation is: $rate = k[A]^2$

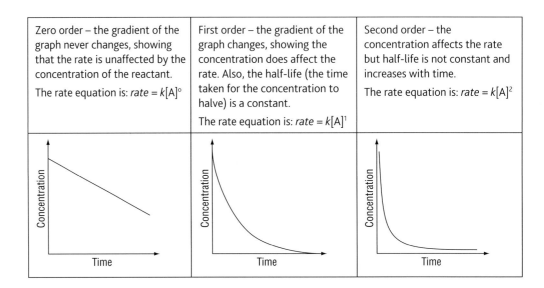

■ WORKED EXAMPLE

The reaction between propanone and iodine can be represented by the equation shown below:

$$H_3C-\overset{\overset{O}{\|}}{C}-CH_3 + I_2 \xrightarrow{\text{H}^+ \text{ catalyst}} I-CH_2-\overset{\overset{O}{\|}}{C}-CH_3 + H^+ + I^-$$

When the concentrations of propanone, H$^+$ ion and iodine are plotted against time, the graphs below are obtained. What is the rate equation for the reaction?

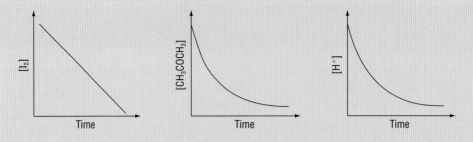

STEP 1 Analyse the graphs and see what category they fit into. The order of reaction with respect to I$_2$ is zero order; for both propanone and H$^+$ ions it is first order.

STEP 2 Put these into the rate equation $rate = k[CH_3COCH_3][H^+]$. Note that $[I_2]^0 = 1$.

Reaction half-life, $t_{1/2}$

The units of half-lives depend on the speed of the reaction. They can be in seconds, minutes, hours, days or years. For example, the decomposition of the gas N_2O_5 has $t_{1/2} = 24.0$ min at 45 °C. So if we start with 0.06 mol dm^{-3} N_2O_5, after 24 min (one half-life) 0.03 mol dm^{-3} remains, so 0.03 mol dm^{-3} has decomposed. After another 24 min, 0.015 mol dm^{-3} remains and 0.015 mol dm^{-3} has decomposed, and so on.

Half-life of a first-order reaction

$t_{1/2}$ is a constant for a first-order reaction, no matter what the concentration of the reactant is.

Look at the concentration–time graph for a first-order reaction. This shape is called an *exponential curve*. Exponential curves always have a constant half-life.

■ WORKED EXAMPLE

How to calculate the half-life – see the curve opposite:

STEP 1 Choose any convenient concentration value (one which can be easily halved!) – here it is 3.2 mol dm^{-3}.

STEP 2 Draw a horizontal line to the curve and then a perpendicular down to the time axis.

STEP 3 Find half the initial concentration value – here it is 1.6 mol dm^{-3} – and find the time again. You now have two time values, and the difference between them is the half-life, $t_{1/2}$. Here it is 10 s.

STEP 4 Find half the concentration again – here it is 0.8 mol dm^{-3} – and repeat the whole process to find the second half-life. Here it is 10 s. If the two $t_{1/2}$ values are the same, the reaction is first order. If they are different, it is not a first-order reaction.

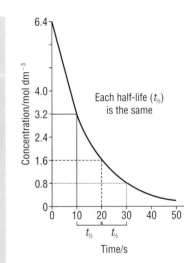

QUICK CHECK QUESTIONS

1   Determine whether the following reaction is first-order for bromine:

$Br_2(aq) + HCOOH (aq) \rightarrow 2Br^-(aq) + 2H^+ (aq) + CO_2(g)$

[Br$_2$]/mol dm^{-3}	Time/min
0.0111	0
0.0081	60
0.0066	120
0.0044	240
0.0020	480
0.0013	600

2   The decomposition of hydrogen peroxide (H_2O_2) to give oxygen and water is first-order with respect to the concentration of hydrogen peroxide.
(a) Give the balanced symbol equation for the reaction.
(b) Give the rate equation for the reaction.
(c) Under certain conditions the half-life of the reaction is five minutes. Draw a sketch graph to show how a 1 mol dm^{-3} solution of hydrogen peroxide decomposes over 25 minutes.

How fast? Rate–concentration graphs

Look at the concentration–time graphs on the previous page. It can be difficult to tell the difference between a first-order and a second-order curve – the second-order curve is steeper, and does not have a constant $t_{1/2}$, but these differences can be difficult to spot. To tell the difference more clearly, we plot *rate–concentration graphs*.

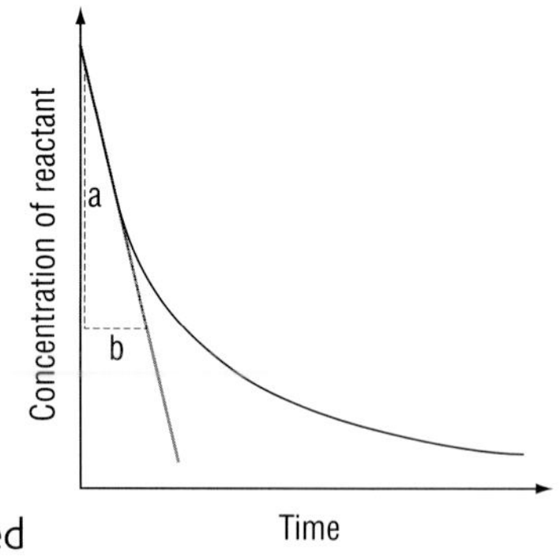

Rate–concentration graphs are obtained by the initial rates method. This means that the **initial rate** is measured in several experiments in which the reactants have different concentrations. The concentration of one reactant at a time is changed while the others remain constant.

It is often more convenient to use the initial rates method than to take concentration and time measurements, especially if it's a slow reaction!

For first- and second-order reactions the initial rate is found by drawing a tangent to the curve at $t = 0$; the gradient gives the rate of the reaction.

Zero order	
The rate equation is: $rate = k[A]^0$ Changing the concentration has no effect on rate.	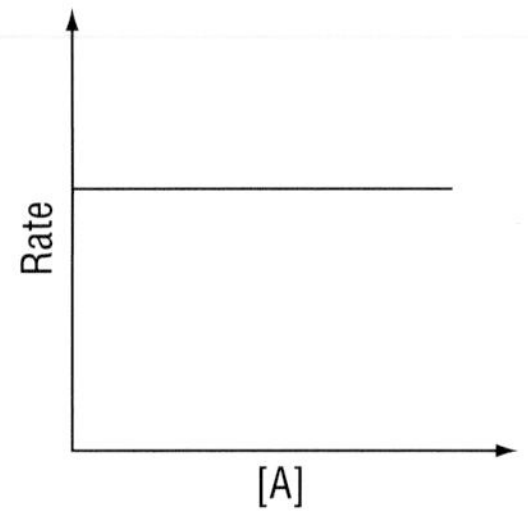
First order	
The general rate equation is: $rate = k[A]^1$ The rate is proportional to the concentration. NOTE k = gradient of line	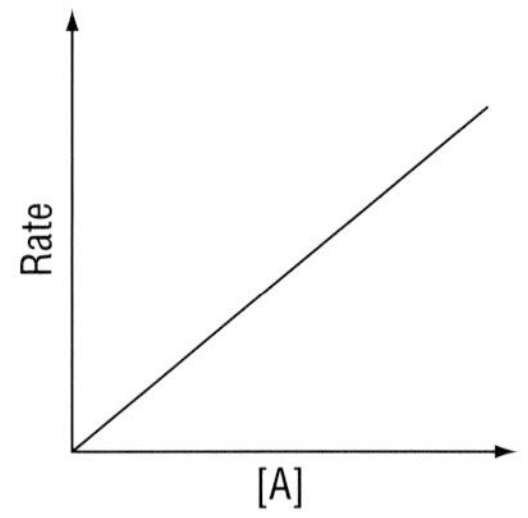
Second order	
The general rate equation is $rate = k[A]^2$	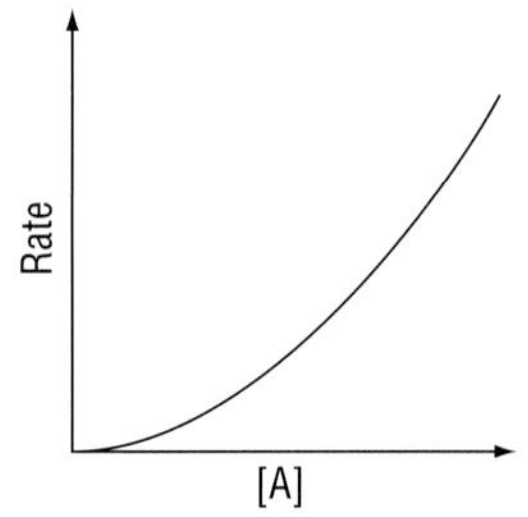

Examiner tip

If you plot rate against (concentration)^2 you get a straight line for a second-order reaction.

✔*Quick check 1*

48

How to calculate the order of a reaction

There are different ways of doing this, which depend on the information you are given.
Follow this map to work out the order:

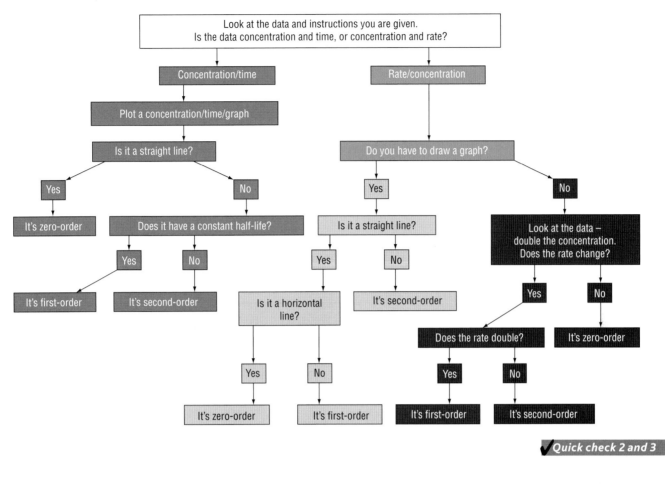

✓ *Quick check 2 and 3*

QUICK CHECK QUESTIONS

1 Describe the initial rates method. Sketch the graph you would obtain for a second-order reaction using this method.

2 Calculate the order of reaction for N_2O_5 in the following reaction:

$$2N_2O_5 (g) \rightarrow 4NO_2(g) + O_2(g)$$

Initial [N_2O_5]/mol dm^{-3}	Initial rate/× 10^{-5} mol dm^{-3} s^{-1}
3.00	3.15
2.51	2.64
1.12	1.18
0.50	0.53

3 Two compounds A and B react as follows:

$$2A + B \rightarrow \text{products}$$

(a) Using the data in the table below, write the rate equation for the reaction between A and B.

Experiment	[A] /mol dm^{-3}	[B] /mol dm^{-3}	Rate /mol dm^{-3} s^{-1}
1	0.001	0.001	1.04×10^{-3}
2	0.002	0.001	2.09×10^{-3}
3	0.001	0.003	1.04×10^{-3}

(b) Draw the graphs to show how rate changes with concentration for both A and B.

49

UNIT 2

How far? The equilibrium law and K_c

✓ *Quick check 1*

Consider a general equilibrium in aqueous solution:

$$aA(aq) + bB(aq) \rightleftharpoons cC(aq) + dD(aq)$$

where A, B, C and D are the reactants, and a, b, c and d are the stoichiometric number of moles indicated in the equation. At any time, once a **dynamic equilibrium** (steady state) has been reached, there will be A, B, C and D present in the reaction at the same time. If there is lots of A and B, then the reaction is not proceeding. If there is lots of C and D, then lots of products have been formed.

The **equilibrium law** states that $K_c = \dfrac{[C]^c[D]^d}{[A]^a[B]^b}$

K_c is the **equilibrium constant** in terms of concentration.

[C], [D], etc are the concentrations *at equilibrium*.

Often these are written as $[C(aq)]^c_{eq}$, $[D(aq)]^c_{eq}$, etc.

The *units of K_c* depend on the equation.

Let's see what this means by writing an equation for the equilibrium constant for this reaction:

$$2SO_2(g) + O_2(g) \rightleftharpoons 2SO_3(g)$$

We simply write the concentration of the product, $[SO_3]$, and raise it to the power of the number of moles shown in the equation, which is two. Divide this by the concentration of the reactants raised to the power of their number of moles as shown in the equation:

$$K_c = \frac{[SO_3]^2}{[SO_2]^2\,[O_2]}$$

The units of K_c

What are the *units of K_c*? Well, the units of concentration are mol dm⁻³, so substitute mol dm⁻³ in the equation for K_c:

$$\text{If } K_c = \frac{[SO_3]^2}{[SO_2]^2\,[O_2]}$$

$$\text{units} = \frac{(\text{mol dm}^{-3})^2}{(\text{mol dm}^{-3})^2\,(\text{mol dm}^{-3})}$$

$$\text{units} = \frac{1}{\text{mol dm}^{-3}}$$

$$= \text{dm}^3\,\text{mol}^{-1}$$

The important thing about the units of K_c is that they are not always the same. Let's look at this reaction:

$$H_2(g) + CO_2(g) \rightleftharpoons CO(g) + H_2O(g)$$

The equation for K_c is:

$$k_c = \frac{[CO]\,[H_2O]}{[H_2]\,[CO_2]}$$

so the units are

$$\frac{(\text{mol dm}^{-3})\,(\text{mol dm}^{-3})}{(\text{mol dm}^{-3})\,(\text{mol dm}^{-3})}$$

which cancels out to *no units*.

You can see how the units for K_c depend on the equation for K_c, which in turn depends on the stoichiometric equation. So always make sure you are using the correct stoichiometric equation when you write the expression for K_c.

 ✓*Quick check 2*

Important facts about K_c

A large value of K_c ($\gg 1$) means a high yield of products at equilibrium.

A small value of K_c ($\ll 1$) means that a high proportion of reactants is present at equilibrium.

> **Examiner tip**
>
> Many organic reactions are equilibrium reactions.

The effect on K_c when conditions are changed

Students often get confused between the change in the *position of the equilibrium* determined by le Chatelier's principle, and K_c.

The main point to remember here is that the value of K_c *only changes if the temperature is changed*. If the *concentration or pressure* is changed, the *position* of the equilibrium may change – remember le Chatelier's principle – but the *value of K_c* stays the same. *Changes in pressure or concentration affect the yield of products but not the value of K_c.*

What happens to K_c when the temperature is changed?

This depends on whether the equilibrium is exothermic or endothermic.

> **Examiner tip**
>
> Le Chatelier's principle states that if an equilibrium is disturbed, the position of the equilibrium will shift to minimise the effect of the change.

✓*Quick check 3*

> **Examiner tip**
>
> The presence or absence of a catalyst has *no effect* on either K_c or the equilibrium yield.

	Exothermic reaction	Endothermic reaction
Temperature raised	K_c decreases	K_c increases
Temperature reduced	K_c increases	K_c decreases

> **Examiner tip**
>
> The terms *exothermic* and *endothermic* used here refer to the thermochemical nature of the *forward reaction*.

QUICK CHECK QUESTIONS

1 Write expressions for K_c, including units, for the following equilibria:
   (a) $PCl_5(g) \rightleftharpoons PCl_3(g) + Cl_2(g)$
   (b) $C_2H_4(g) + H_2O(g) \rightleftharpoons C_2H_5OH(g)$
   (c) $H_2(g) + Br_2(g) \rightleftharpoons 2HBr(g)$

2 Fill in the gaps (a)–(l) in the table below.

Equilibrium	ΔH + or –	Change made	Effect on yield of products	Effect on K_c
$PCl_5(g) \rightleftharpoons PCl_3(g) + Cl_2(g)$	+	Increase pressure	(a)	(b)
$PCl_5(g) \rightleftharpoons PCl_3(g) + Cl_2(g)$	+	Increase temperature	(c)	(d)
$C_2H_4(g) + H_2O(g) \rightleftharpoons C_2H_5OH(g)$	–	Increase pressure	(e)	(f)
$C_2H_4(g) + H_2O(g) \rightleftharpoons C_2H_5OH(g)$	–	Increase temperature	(g)	(h)
$H_2(g) + Br_2(g) \rightleftharpoons 2HBr(g)$	–	Increase pressure	(i)	(j)
$H_2(g) + Br_2(g) \rightleftharpoons 2HBr(g)$	–	Increase temperature	(k)	(l)

Module 1

How far? How to calculate the value of K_c

You need to be able to calculate the numerical value of K_c. There are two different types of approach, depending on the information you are given in the question.

1 If you are given the *equilibrium* concentrations

The equation for K_c uses the concentrations at equilibrium. If you are given these, simply substitute into the K_c equation:

> ■ **WORKED EXAMPLE**
>
> Calculate K_c for
>
> $$H_2(g) + CO_2(g) \rightleftharpoons CO(g) + H_2O(g)$$
>
> given $[H_2]_{eq} = 0.530$ mol dm^{-3}, $[CO_2]_{eq} = 0.805$ mol dm^{-3},
>
> $$[CO]_{eq} = 9.47 \text{ mol dm}^{-3}, [H_2O]_{eq} = 9.47 \text{ mol dm}^{-3}$$
>
> **STEP 1** Write out the expression for K_c from the stoichiometric equation.
>
> $$K_c = \frac{[CO]\,[H_2O]}{[H_2]\,[CO_2]}$$
>
> **STEP 2** Substitute with the equilibrium concentrations you are given.
>
> $$K_c = \frac{[CO]\,[H_2O]}{[H_2]\,[CO_2]} = \frac{9.47 \times 9.47}{0.53 \times 0.805} = 210 \text{ (no units)}$$

Examiner tip

Imagine you were *not* given $[H_2O]_{eq}$. What would you do? Well, the stoichiometric equation tells you that at equilibrium the number of moles of H_2O and the number of moles of CO are the same. So if $[CO] = 9.47$ mol dm^{-3} then $[H_2O] = 9.47$ mol dm^{-3}.

2 If you are given the *initial* concentrations

What if the equilibrium concentrations of the reactants are unknown? You can calculate them from *initial* amounts of the reactants and the *equilibrium concentration of the product*.

If you know how many moles of product are present *at equilibrium*, the stoichiometric equation will tell you how many moles of reactant are present *at equilibrium*.

In another type of question you could be given the percentage conversion of reactant to product.

> ■ **WORKED EXAMPLE**
>
> Calculate K_c for the equilibrium
>
> $$2SO_2(g) + O_2(g) \rightleftharpoons 2SO_3(g)$$
>
> if 2.00 mol of SO_2 and 1.00 mol of O_2 were introduced into the equilibrium vessel (volume 20 dm^3) and 1.80 mol of SO_3 were formed at equilibrium.
>
> **STEP 1** Construct the expression for K_c:
>
> $$K_c = \frac{[SO_3]^2}{[SO_2]^2\,[O_2]}$$

✓ *Quick check 1*

How far? How to calculate the value of K_c

Module 1

STEP 2 Make up a table based on the stoichiometric equation. You are going to need the *initial number of moles* and the *equilibrium number of moles*. Finally you are going to need *equilibrium concentrations in mol dm⁻³*.

	$2SO_2$	$+ O_2$	$\rightleftharpoons$	$2SO_3$
Initial amounts/mol				
Equilibrium amounts/mol				
Equilibrium concentration/mol dm⁻³				

STEP 3 Put the values you know under the appropriate species in the equation.

	$2SO_2$	$+ O_2$	$\rightleftharpoons$	$2SO_3$
Initial amounts/mol	2.00 mol	1.00 mol		0
Equilibrium amounts/mol				1.80 mol
Equilibrium concentration/mol dm⁻³				

STEP 4 Now you can put in the number of moles of the other species and work out equilibrium concentrations.

	$2SO_2$	$+ O_2$	$\rightleftharpoons$	$2SO_3$
Initial amounts /mol	2.00 mol	1.00 mol		0
Equilibrium amounts/mol	2.00–1.80 = 0.200 mol since for every 1 mol of SO_3 formed, 1 mol of SO_2 must have reacted.	1–0.900 = 0.100 mol since for every 2 mol of SO_2 that react, 1 mol of O_2 has reacted.		1.80 mol
Equilibrium concentration /mol dm⁻³ (divide amount in mol by volume in dm³)	0.200/20.0 = 0.0100 mol dm⁻³	0.100/20.0 = 0.00500 mol dm⁻³		1.80/20.0 = 0.0900 mol dm⁻³

STEP 5 Substitute these values into the expression for K_c and calculate the units.

$$K_c = \frac{(0.09)^2}{(0.01)^2(0.005)} = 1.62 \times 10^3$$

$$\frac{(mol\ dm^{-3})^2}{(mol\ dm^{-3})^2 \times mol\ dm^{-3}} = \frac{1}{mol\ dm^{-3}} = dm^3\ mol^{-1}$$

✔ *Quick check 1 and 2*

QUICK CHECK QUESTIONS

1 0.204 mol $CH_3COOC_2H_5$ and 0.645 mol H_2O were introduced into a flask, acidified and left until a dynamic equilibrium (steady state) was reached. A sample was then analysed, and 0.114 mol of CH_3COOH was found. Calculate K_c for the reaction.

$CH_3COOC_2H_5(l) + H_2O(l) \rightleftharpoons CH_3COOH(l) + C_2H_5OH(l)$

2 Under certain conditions, when hydrogen and iodine react in a 1:1 molar ratio, 80% of the iodine reacts to form hydrogen iodide (HI). Calculate K_c and give the units.

Acids and bases

The chemistry of acids and bases is an important area. For example, the relative strengths of acids influence the formation of nitronium ions in the nitration of benzene, and the understanding of pH and buffers (see page 60) is essential in biology.

The Brønsted–Lowry theory of acids and bases

- Acids are **proton** (H^+ ion) **donors** whilst bases are **proton acceptors**.
- In an acid–base reaction, the acid donates a proton to form a **conjugate base** and the base accepts a proton to form a **conjugate acid**.
- Bases must always have lone-pair electrons to form a dative covalent bond with the proton.
- An **acid–base pair** (a conjugate acid and conjugate base) is a set of two species that transform into each other by gain or loss of a proton.

For example, consider the dissociation of ethanoic acid in water:

$$CH_3COOH(aq) + H_2O(l) \rightleftharpoons CH_3COO^-(aq) + H_3O^+(aq)$$

In the forward direction the conjugate acid CH_3COOH releases a proton to form its conjugate base CH_3COO^-. In the reverse reaction the conjugate base CH_3COO^- accepts the proton to form its conjugate acid CH_3COOH. Thus there are two acid–base pairs:.

$$CH_3COOH(aq) + H_2O(l) \rightleftharpoons CH_3COO^-(aq) + H_3O^+(aq)$$

    Acid 1   Base 1
          Base 2   Acid 2

- The H_3O^+ ion is called the hydronium ion and is the actual ion found in aqueous solutions of acids. The H^+ ion cannot really exist. It is not stable because it does not have the two electrons necessary for stability, and it immediately attracts water molecules because of its small size and dense charge. However, we can represent it as $H^+(aq)$.
- Water can also act as an acid when it is with a stronger proton acceptor such as ammonia.

$$NH_3(aq) + H_2O(l) \rightleftharpoons NH_4^+(aq) + OH^-(aq)$$

    Base 1   Acid 1
          Acid 2   Base 2

- The strength of an acid depends on its ability to donate protons. The stronger the acid, the stronger its ability to donate protons.
- The strength of a base depends on its ability to accept protons. The stronger the base, the stronger its ability to accept protons.

The reaction between concentrated sulfuric acid and concentrated nitric acid is an acid–base reaction because the sulfuric acid is the stronger proton donor.

$$HNO_3(l) + H_2SO_4(l) \rightleftharpoons H_2NO_3^+(l) + HSO_4^-(l)$$

    Base 1   Acid 1
          Acid 2   Base 2

- Water can act as an acid (able to donate protons) and as a base (able to accept protons).

Key words

- proton donor
- proton acceptor
- conjugate base
- conjugate acid
- acid–base pair

✓ Quick check 1

Examiner tip

The arrows can be written with one longer than the other to represent an equilibrium that is well over to one side.

✓ Quick check 1

✓ Quick check 2

Examiner tip

Acids such as H_2SO_4 which have two protons that they can donate are called diprotic acids. HCl has one proton to donate and is monoprotic.

✓ Quick check 3

The reactions of acids

At AS level you had to learn about the reactions of acids in aqueous solution. We know that in aqueous solution *all* acids produce the hydronium (H_3O^+) ion and therefore they have reactions that are characteristic of this ion. In equations we can write this as the $H^+(aq)$ ion. The table below summarises the reactions and gives the ionic equations for the reactions taking place.

✓*Quick check 2*

Examiner tip

The ionic equations are applicable to all acids.

Reaction	Symbol and Ionic equation
With reactive metals such as magnesium to give **hydrogen gas** and **a salt**	e.g. $2HCl(aq) + Mg(s) \rightarrow H_2(g) + MgCl_2(aq)$ $2H^+(aq) + Mg(s) \rightarrow H_2(g) + Mg^{2+}(aq)$
With carbonates to give **water, carbon dioxide** and **a salt**	e.g. $2HCl(aq) + Na_2CO_3(aq) \rightarrow H_2O(l) + CO_2(g) + 2NaCl(aq)$ $2H^+(aq) + CO_3^{2-}(aq) \rightarrow H_2O(l) + CO_2(g)$
With bases to give a **salt** and **water**	e.g. $2HCl(aq) + MgO(s) \rightarrow H_2O(l) + MgCl_2(aq)$ $2H^+(aq) + Mg^{2+} O^{2-}(s) \rightarrow H_2O(l) + Mg^{2+}(aq)$
With alkalis (bases soluble in water) to give a **salt** and **water**	e.g. $HCl(aq) + NaOH(aq) \rightarrow H_2O(l) + Na^+Cl(aq)$ $H^+(aq) + OH^-(aq) \rightarrow H_2O(l)$

✓*Quick check 4*

QUICK CHECK QUESTIONS

1 For the acid–base equilibria shown below, identify the acid–base pairs.
  (a) $HClO_4 + HNO_3 \rightleftharpoons ClO_4^- + H_2NO_3^+$
  (b) $HBr + H_2O \rightleftharpoons H_3O^+ + Br^-$
  (c) $CH_3NH_2 + H_2O \rightleftharpoons CH_3NH_3^+ + OH^-$

2 (a) What is the name given to the H_3O^+ ion?
  (b) From question **1** above, give one reaction in which water acts as a base and one in which it acts as an acid.

3 $HClO_4$ and HNO_3 are both acids in aqueous solution.
  (a) Give the equations for their reactions with water.
  (b) Using the relevant equation in question **1(a)**, explain which of these two compounds is the stronger proton donor.

4 Give the ionic equations for the reactions of the following substances with sulfuric acid.
  (a) Magnesium metal
  (b) Copper carbonate
  (c) Calcium oxide
  (d) Potassium hydroxide.

Key words

• pH scale

Examiner tip

p is shorthand for $-\log_{10}$.

✓*Quick check 1*

Examiner tip

Remember, in aqueous solution the term $H^+(aq)$ is used instead of $[H_3O^+(aq)]$.

The pH scale

The concentrations of aqueous hydrogen ions vary widely and the range would be impossible to represent on a linear scale. To overcome this problem a logarithmic scale is used, called the **pH scale**.

On this scale

$$pH = -\log_{10}[H^+(aq)]$$

It follows that $[H^+(aq)]$ can be written in terms of the pH.

$$[H^+(aq)] = 10^{-pH}$$

To calculate pH you simply have to use your calculator keys correctly. If you can do that, then this section isn't difficult.

■ **WORKED EXAMPLE 1**

What is the pH of a solution with $[H^+(aq)]$ equal to 3×10^{-4} mol dm^{-3}?

The exact procedure depends on your calculator and you should consult your manual.

Typical procedures are as follows:

STEP 1 Press the [(−)] key (on some calculators this is +/−), then [log] 3 [EXP] [(−)] 4.

STEP 2 Press [=] . You should get 3.52. Remember, keep it to two decimal places.

STEP 3 Write your answer: pH = 3.52.

■ **WORKED EXAMPLE 2**

What is the $[H^+(aq)]$ if the pH is 7.2?

To find $[H^+(aq)]$ we use the formula $[H^+] = 10^{-pH}$.

STEP 1 Find the [10^x] key and press it. **NOTE:** The 10^x key is usually [SHIFT][log].

STEP 2 The procedure for this number is as follows [SHIFT][log][−] 7.2.

STEP 3 Press your [=] key and you get the answer: 6.31×10^{-8} mol dm^{-3}.

✓*Quick check 2*

The ionisation of water

• Water ionises as follows: $H_2O(l) \rightleftharpoons H^+(aq) + OH^-(aq)$.
  The equilibrium constant may therefore be written as shown below.

$$K = \frac{[H^+(aq)]\,[OH^-(aq)]}{[H_2O(l)]}$$

Remember this

$K_w = [H^+(aq)]_{eq}\,[OH^-(aq)]_{eq}$ mol^2 dm^{-6}

• At 25 °C this equilibrium lies well to the left and the equilibrium concentrations of $[H^+(aq)]$ and $[OH^-(aq)]$ are both equal to 10^{-7} mol dm^{-3}. Therefore $[H_2O(l)]_{eqm}$ is virtually unchanged from that of undissociated water and can be treated as constant. A new equilibrium constant K_w is now used. It can be represented by the equation on the left.

- At 25 °C $K_w = 1.00 \times 10^{-14}$ mol² dm⁻⁶
  Since $[H^+(aq)] = [OH^-(aq)]$, then $K_w = [H^+(aq)]^2$
  $[H^+(aq)]^2 = 1 \times 10^{-14}$ mol² dm⁻⁶

  $$[H^+(aq)] = \sqrt{1 \times 10^{-14} \text{ mol}^2 \text{ dm}^{-6}} = 1 \times 10^{-7} \text{ mol dm}^{-3}$$

- When we work out the pH for this $[H^+(aq)]$ we get a value of 7, which explains why for all these years you have associated this value with a neutral pH.

The pH values of aqueous solutions

This is a section where we have to assume you know the expression for pH, and once again you must use your calculator.

The pH values of solutions of strong acids

- The important point here is that strong acids are fully dissociated in aqueous solution and therefore we can assume 100% dissociation.

- For example, 0.1 mol dm⁻³ HCl dissociates as follows:

  $HCl(aq) \rightarrow H^+(aq) + Cl^-(aq)$

- For 100% dissociation every HCl gives one $H^+(aq)$.

Therefore $[H^+(aq)] = [HCl(aq)] = 0.1$ mol dm⁻³.

Therefore pH = $-\log_{10} 0.1 = 1$.

The pH values of solutions of strong bases

- For strong bases the assumption of 100% dissociation is still valid.

- For example, NaOH dissociates as follows:

  $NaOH(aq) \rightarrow Na^+(aq) + OH^-(aq)$

- Therefore, if we have a solution of 0.02 mol dm⁻³ NaOH, then every NaOH gives an $OH^-(aq)$ ion . Therefore $[OH^-(aq)] = [NaOH(aq)] = 0.02$ mol dm⁻³.

- But pH is calculated on the basis of $[H^+(aq)]$ and therefore we have to find $[H^+(aq)]$ using our value of $[OH^-(aq)]$. To do this we use the relationship:

  $K_w = [H^+(aq)][OH^-(aq)] = 1 \times 10^{-14}$ mol² dm⁻⁶

Substituting in our value for $[OH^-(aq)]$ we have:

  $1 \times 10^{-14} = [H^+(aq)] \times 0.02$

  $[H^+(aq)] = 1 \times 10^{-14}/0.02 = 5 \times 10^{-13}$ mol dm⁻³

Therefore pH = $-\log_{10} 5 \times 10^{-13} = 12.3$.

QUICK CHECK QUESTIONS

1 Calculate the pH values of solutions with the following $[H^+(aq)]$ values:
   (a) 0.01 mol dm⁻³
   (b) 2×10^{-9} mol dm⁻³
   (c) 3×10^{-4} mol dm⁻³.

2 Calculate the concentration of $H^+(aq)$ ions for solutions with the following pH values:
   (a) 2.5   (b) 7.4
   (c) 10.5   (d) 13.6.

3 Calculate the pH values of the following solutions:
   (a) HCl
      (i) 0.010 mol dm⁻³
      (ii) 3×10^{-5} mol dm⁻³.
   (b) NaOH
      (i) 0.010 mol dm⁻³
      (ii) 3×10^{-5} mol dm⁻³.

UNIT 2

The chemistry of weak acids

Module 1

Key words

- partial dissociation
- acid dissociation constant

- Monobasic (monoprotic) acids (one proton produced per molecule) can be represented as HA, where H is the hydrogen.
- Weak acids are poor proton donors and are only **partially dissociated** in aqueous solution. Therefore the dissociation is not 100% and $[H^+(aq)]$ cannot be calculated directly from the concentration of the acid.
- The equilibrium between a weak acid and its constituent ions in aqueous solution is generally written:

$$HA(aq) \rightleftharpoons H^+(aq) + A^-(aq)$$

- The equilibrium constant for this reaction is called the **acid dissociation constant** (**K_a**). The expression for K_a is shown below.

$$K_a = \frac{[H^+(aq)]\,[A^-(aq)]}{[HA(aq)]}$$

Examiner tip

The magnitude of K_a is a measure of how much the acid ionises. Therefore the higher the value of K_a the stronger the acid.

Examiner tip

The units of K_a are $mol\,dm^{-3}$ because the units in the expression are

$$\frac{mol\,dm^{-3} \times mol\,dm^{-3}}{mol\,dm^{-3}}$$

$= mol\,dm^{-3}$

The meaning of the term pK_a

$pK_a = -\log_{10}K_a$. For example, if $K_a = 4.8 \times 10^{-5}\,mol\,dm^{-3}$,

$pK_a = -\log_{10}4.8 \times 10^{-5} = 4.32$.

RULE: The higher the value of pK_a, the weaker the acid.

Quick check 1

■ WORKED EXAMPLE

Which is the stronger of the two acids, ethanoic acid ($K_a = 1.7 \times 10^{-5}\,mol\,dm^{-3}$) or chloric(I) acid ($K_a = 3.7 \times 10^{-8}\,mol\,dm^{-3}$)?

Calculate the p*K*a values:

Ethanoic acid $pK_a = -\log_{10}1.7 \times 10^{-5} = 4.77$

Chloric(I) acid $pK_a = -\log_{10}3.7 \times 10^{-8} = 7.43$

The weaker of the two acids is therefore chloric(I) acid.

Finding K_a from pK_a

As with pH, the value of K_a can be deduced from pK_a by using the relationship $K_a = 10^{-pK_a}$.

Examiner tip

This is the opposite way round to the rule for K_a.

■ WORKED EXAMPLE

The pK_a of phenol is 9.9. What is its K_a?

$K_a = 10^{-pK_a} = 10^{-9.9} = 1.26 \times 10^{-10}\,mol\,dm^{-3}$.

There are two types of question which involve K_a and $[H^+(aq)]$.

Type 1 Finding K_a from pH

■ WORKED EXAMPLE

What is K_a for a weak acid if a solution of the acid with concentration 0.0100 mol dm^{-3} has a pH of 5.50?

STEP 1 Calculate [H$^+$(aq)]. In this case the pH is 5.50.*
Therefore **[H$^+$(aq)] = 10$^{-5.50}$ = 3.16 × 10^{-6} mol dm^{-3}**

STEP 2 Since there are equal numbers of H$^+$(aq) and A$^-$(aq) produced in the dissociation, we can assume that **[H$^+$(aq)] = [A$^-$(aq)]**.
Therefore **[H$^+$(aq)][A$^-$(aq)] = [H$^+$(aq)]^2**
In this case **[H$^+$(aq)]^2 = (3.16 × 10^{-6})^2 = 9.99 × 10^{-12} mol^2 dm^{-3}**

STEP 3 Here we make the assumption that *[HA(aq)] at equilibrium is virtually the same as it was at the start*, because as a weak acid so little of HA ionises.

[HA(aq)]$_{eqm}$ ≈ [HA(aq)]$_{start}$ = 0.0100 mol dm^{-3}

STEP 4 Substitute these values into the expression for K_a and we have:
$$K_a = \frac{9.99 \times 10^{-12}}{0.0100} = 9.99 \times 10^{-10} \text{ mol dm}^{-3}$$
Note: The low value for K_a emphasises that this is a weak acid.

> **Examiner tip**
>
> * If this was a strong acid and 100% dissociated, [H$^+$(aq)] would be 0.01 mol dm^{-3} and the pH would be 2.

✓*Quick check 2*

Type 2 Finding pH from K_a

■ WORKED EXAMPLE

The value of K_a for a weak acid is 4.80 × 10^{-5} mol dm^{-3}. What is the pH of a solution of the acid with a concentration 0.100 mol dm^{-3}?

STEP 1 Make the same assumptions as before,
i.e. **[H$^+$(aq)] = [A$^-$(aq)]** and **[HA(aq)]$_{eqm}$ ≈ [HA(aq)]$_{start}$**
Therefore K_a = [H$^+$(aq)]^2/[HA(aq)]$_{start}$
Therefore [H$^+$(aq)]^2 = K_a × [HA(aq)]$_{start}$

STEP 2 Substitute in these values.
Therefore **[H$^+$(aq)]^2 = 4.80 × 10^{-5} × 0.100 = 4.80 × 10^{-6}**
Therefore **[H$^+$(aq)] = √(4.80 × 10^{-6}) = 2.19 × 10^{-3} mol dm^{-3}**

STEP 3 Calculate pH from [H$^+$(aq)]. pH = −log$_{10}$2.19 × 10^{-3} = 2.66

Note: The question may give you the pK_a and not K_a and ask you to find the pH. If this is the case, simply calculate K_a using K_a = 10^{-pKa} and carry on as in the example.

✓*Quick check 3*

QUICK CHECK QUESTIONS

1 The K_a values of two weak acids HA and HB are as follows:

K_a (HA) = 2.9 × 10^{-3} K_a (HB) = 4.5 × 10^{-6}

(a) Calculate pK_a of (i) HA and (ii) HB.
(b) Which is the stronger of the two acids?
(c) Write an equation for the reaction between HA and HB (refer to the Brønsted–Lowry theory).

2 The pH of a 0.0100 mol dm^{-3} solution of butanoic acid is 3.41. What is the value of K_a for the acid?

3 The pK_a of silicic acid is 9.9.
(a) What is the value of K_a?
(b) What is the pH of a 0.001 mol dm^{-3} solution of the acid?

Buffers – how they work and their pH

Key words

- buffers
- weak acids
- conjugate bases

A **buffer** solution is one that minimises pH changes when small amounts of acid or base are added. Buffers are important solutions, especially in biological systems. A good understanding of how they work is essential if you are to use them properly.

How buffer solutions work

The type of buffer solution you need to know about is that consisting of a **weak acid** and its **conjugate base** pair, for example ethanoic acid and ethanoate ions (CH_3COOH/ CH_3COO^-).

In this first example, the CH_3COO^- ions are supplied by a solution of the sodium salt, sodium ethanoate, which is fully dissociated.

This type of buffer will give a pH less than 7.

How do buffer solutions resist changes in pH?

Example: ethanoic acid and ethanoate ions

The equilibrium is: $CH_3COOH(aq) \rightleftharpoons H^+(aq) + CH_3COO^-(aq)$

The stable pH of the acidic buffer is maintained by:

- a high $[CH_3COO^-(aq)]$, which mops up added $H^+(aq)$ by forming the weak (and therefore largely undissociated) acid, ethanoic acid
- a high $[CH_3COOH(aq)]$, which supplies $H^+(aq)$ to mop up added $OH^-(aq)$ ions by neutralising them. As the $[OH^-(aq)]$ decreases, the equilibrium is disturbed. The CH_3COOH dissociates until K_a is restored to its proper value and therefore the $[H^+(aq)]$, and hence the pH, is back to near its previous value.

Examiner tip

Remember the neutralisation reaction is $H^+(aq) + OH^-(aq) \rightarrow H_2O(l)$

✔ *Quick check 1*

Calculating the pH of a buffer solution

The equation used is a rearrangement of the equation for the acid dissociation constant of a weak acid.

$$[H^+(aq)] = K_a \times \frac{[acid]}{[salt]}$$

■ WORKED EXAMPLE

What is the pH of a solution 0.02 mol dm^{-3} with respect to ethanoic acid ($K_a = 1.7 \times 10^{-5}$ mol dm^{-3}) and 0.05 mol dm^{-3} with respect to sodium ethanoate?

STEP 1 Substituting the values for K_a, [acid] and [salt] into the equation we have:

$$[H^+(aq)] = 1.7 \times 10^{-5} \times (0.02/0.05) = 6.8 \times 10^{-6} \text{ mol dm}^{-3}$$

STEP 2 Calculate pH from $[H^+(aq)]$

$$pH = -\log_{10} 6.8 \times 10^{-6} = 5.17$$

An alternative equation for the pH of a buffer is:

$$pH = pK_a + \log_{10} \frac{[salt]}{[acid]}$$

■ WORKED EXAMPLE

What is the pH of a solution containing 100 cm³ of 0.01 mol dm⁻³ ethanoic acid ($K_a = 1.7 \times 10^{-5}$ mol dm⁻³) and 200 cm³ of 0.02 mol dm⁻³ sodium ethanoate?

STEP 1 Calculate the final concentrations of the ethanoic acid and sodium ethanoate.

The final volume is 300 cm³.

Therefore the concentration of the ethanoic acid = (100/300) × 0.01
$$= 0.00333 \text{ mol dm}^{-3}.$$

The concentration of sodium ethanoate = (200/300) × 0.02 = 0.0133 mol dm⁻³.

STEP 2 $\dfrac{[\text{salt}]}{[\text{acid}]} = (0.0133/0.0033) = 4.00$

Therefore $\log_{10} \dfrac{[\text{salt}]}{[\text{acid}]} = \log_{10} 4 = 0.602.$

STEP 3 $pK_a = -\log_{10} K_a = 4.77.$

STEP 4 Substituting these values in the equation

$$pH = 4.77 + 0.602 = 5.37.$$

✔ *Quick check 1*

Buffers in the body

Relatively small changes in pH mean large changes in [H⁺(aq)]. For example, a change of blood pH from 7.4 to 6.8 means that the [H⁺(aq)] has increased fourfold!

The pH of blood plasma has to be kept between 7.35 and 7.45. Haemoglobin acts as a buffer but if its buffering capacity is exceeded then the H_2CO_3/HCO_3^- system acts as a buffer.

The equilibrium is as follows: $H_2CO_3 \text{ (aq)} \rightleftharpoons HCO_3^-\text{(aq)} + H^+\text{(aq)}$

If acidosis occurs, i.e. [H⁺] increases (a low pH caused by excessive exercise), then the equilibrium is disturbed. The H⁺(aq) ions combine with the HCO₃⁻(aq) ions to form carbonic acid, H_2CO_3, thus lowering the [H⁺(aq)] and hence raising the blood pH. This then forms CO_2 gas and water. The CO_2 is breathed out and the blood pH is returned to normal.

✔ *Quick check 3*

QUICK CHECK QUESTIONS

1   Explain how a solution of sodium ethanoate and ethanoic acid can function as a buffer.

2   A solution of 100 cm³ of hydrofluoric acid (0.03 mol dm⁻³) and 50 cm³ of sodium fluoride (0.03 mol dm⁻³) were mixed together.

   (K_a of hydrofluoric acid = 5.6 × 10⁻⁴ mol dm⁻³)

   (a) What is the final concentration of the acid?
   (b) What is the final concentration of the sodium fluoride?
   (c) Calculate the pH of the buffer solution formed.
   (d) Explain how you think a mixture of HF and NaF may act as a buffer solution.

3   Why is saline solution containing HCO₃⁻(aq) given to a person suffering from acidosis?

Indicators, acid–base titration curves and neutralisation

The theory of this section is an extension of the previous sections, so make sure you understand terms like pH and pK_a.

Key words

- indicators
- weak acids
- weak bases
- pH range
- neutralisation
- enthalpy change

Indicators

- **Indicators** are either **weak acids** or **weak bases**.
- The equilibrium for an indicator HI_n (aq) that is a weak acid may be written as:

$$\text{HIn(aq)} \rightleftharpoons \text{H}^+\text{(aq)} + \text{In}^-\text{(aq)}$$

  colour I   colour II

  The **pH range** over which an indicator changes colour is approximately equal to p$K_a \pm 1$.

- Therefore, for HIn at a pH at least 1 pH unit below pK_a the colour of the indicator is colour I. At pH values at least 1 pH unit greater than pK_a it has the colour II.

- An indicator can be used for a titration if the pH range over which it changes colour coincides with a rapid pH change in the titration (see titration curves below).

- Two common indicators are phenolphthalein (abbreviated to PP; p$K_a = 9.3$) and methyl orange (abbreviated to MO; p$K_a = 3.7$). The diagram below shows the pH ranges of these two indicators.

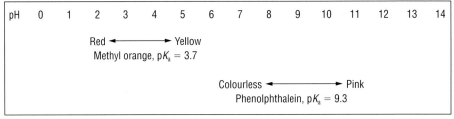

✔ *Quick check 1*

Titration curves

For all the titrations below and on the next page, 0.1 mol dm^{-3} alkali (base) is added to 25 cm^3 of 0.1 mol dm^{-3} acid.

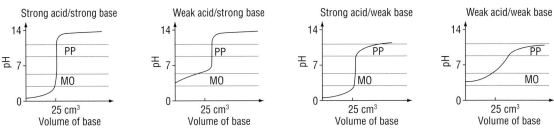

Choice of indicator

Titration	Which indicator	Reasons for choice
Strong acid/strong base	MO or PP	There is a rapid pH change over the colour range for either indicator.
Weak acid/strong base	PP	The rapid pH change is over the range for PP but not for MO.
Strong acid/weak base	MO	The rapid pH change is over the range for MO but not for PP.
Weak acid/weak base	Neither	There is no rapid pH change in any region of the titration curve.

The pK_a values of different indicators are given in the table below:

Indicator	pK_a	Colour change acid $\rightleftharpoons$ alkaline
Congo red	4.0	Violet $\rightleftharpoons$ red
Methyl red	5.1	Red $\rightleftharpoons$ yellow
Thymol blue	8.9	Yellow $\rightleftharpoons$ blue
Thymolphthalein	9.7	Colourless $\rightleftharpoons$ blue

✓*Quick check 1 and 2*

Module 1

Neutralisation

- The **enthalpy change of neutralisation** is the enthalpy change when an aqueous acid is neutralised by an aqueous base to form one mole of $H_2O(l)$ under standard conditions.

- The reaction between all acids and alkalis is $H^+(aq) + OH^-(aq) \rightarrow H_2O(l)$.

- Since all strong acids and alkalis are completely dissociated, the enthalpy change of neutralisation is the same. Therefore when hydrochloric acid is neutralised by sodium hydroxide and nitric acid is neutralised by potassium hydroxide, the enthalpy change of neutralisation is the same.

■ WORKED EXAMPLE

When 50.0 cm³ of 2.00 mol dm⁻³ hydrochloric acid reacts with 50.0 cm³ of 2.00 mol dm⁻³ sodium hydroxide, the temperature increase is 13.8 °C. What is the enthalpy change of neutralisation for the reaction?

STEP 1 Calculate the enthalpy change.

Use the equation:

Heat gained = total mass of water $\times c \times \Delta t$

= 100 × 4.18 × 13.8 = 5768.4 J = 5.77 kJ

STEP 2 Calculate the amount in mol of H^+ ions.

Amount in mol = concentration × volume in dm³

= 2.00 × 0.0500 mol = 0.100 mol

STEP 3 Calculate the enthalpy change per mol of H^+ ions.

0.100 mol give an enthalpy change of –5.77 kJ.

Therefore ΔH per mol = –57.7 kJ = ΔH_{neut}.

✓*Quick check 3*

QUICK CHECK QUESTIONS

1 What are the colours of the indicators in the above table at the following pH values?
   (a) pH 1.0
   (b) pH 7.0
   (c) pH 10.

2 Which of the indicators could you use for the following titrations?
   (a) Strong acid/strong base
   (b) Weak acid/strong base
   (c) Strong acid/weak base
   (d) Weak acid/weak base.

3 When 100.0 cm³ of 2 mol dm⁻³ nitric acid reacts with 100.0 cm³ of 2.00 mol dm⁻³ sodium hydroxide, the temperature change is 13.8 °C. What is the enthalpy change of neutralisation for the reaction?

Lattice enthalpy

Key words

- lattice enthalpy
- exothermic
- endothermic

Examiner tip

Don't forget the state symbols in this definition, especially the gaseous ions on the left.

You have already studied some enthalpy changes at AS level. The lattice enthalpy is a particular type of enthalpy change for ionic solids only. It is the energy released when gaseous ions form a solid ionic lattice.

The lattice enthalpy, $\Delta H^{\ominus}_{latt}$, is the enthalpy change when 1 mole of an ionic compound is formed from its gaseous ions under standard conditions (298 K, 100 kPa).

For example: $M^+(g) + X^-(g) \rightarrow MX(s)$

Here is a checklist of things about the lattice enthalpy you must know. It is:

- always an **exothermic** change
- represented by the symbol $\Delta H^{\ominus}_{latt}$
- a measure of the strength of the ionic lattice in the compound – the ionic bond strength. A large exothermic value for the lattice enthalpy means a large electrostatic attraction between the oppositely charged ions in the lattice.
- not directly measurable by experiment.

Trends in lattice enthalpy

The *size of the ions* and the *charge on the ions* affect the value of the lattice enthalpy. There are two trends in lattice enthalpy values that you need to know.

The lattice enthalpy:

- is *more exothermic* if the *ionic charge increases*, because an increased charge density on the ions means a stronger attraction between them and so a stronger lattice.
- is *more exothermic* if the *ionic radius decreases*, since if the charge remains the same but the ionic radius decreases then the charge density increases, leading to increased attraction between the ions in the ionic lattice.
- NOTE: we always say that the lattice enthalpy is more exothermic or more negative rather than larger, because if we say larger it could mean more positive, which is **endothermic**.

The *more exothermic* the enthalpy change is, the stronger the ionic bond. So if we compare the lattice enthalpies of sodium chloride and sodium iodide, we can see which has the stronger electrostatic attraction between ions:

✓ *Quick check 1 and 3*

$$\Delta H^{\ominus}_{latt} \text{ (NaCl)} = -781 \text{ kJ mol}^{-1} \quad \Delta H^{\ominus}_{latt}\text{(NaI)} = -699 \text{ kJ mol}^{-1}$$

NaCl has the more exothermic lattice enthalpy, so has the stronger ionic bonding.

This is because the Cl^- ion is smaller than the I^- ion and therefore has a higher charge density. It therefore attracts the sodium ion more strongly than does the iodide ion, leading to a more exothermic lattice energy for NaCl.

How to write lattice enthalpy equations

You need to be able to write an equation for the reaction that describes a lattice enthalpy. Look again at the definition of lattice enthalpy – when you write an equation representing a lattice enthalpy, remember that only *one mole* of the solid is formed. Balance the rest of the equation to fit in with this.

State symbols are important here, as gaseous ions react to form a solid. This reaction cannot be done in real life, but we still like to know the value of the lattice enthalpy because it gives useful information about the strength of the ionic lattice.

So, the equation representing the lattice enthalpy of sodium chloride is:

$$Na^+(g) + Cl^-(g) \rightarrow NaCl(s)$$

and for the lattice enthalpy of magnesium oxide the equation is:

$$Mg^{2+}(g) + O^{2-}(g) \rightarrow MgO(s)$$

■ WORKED EXAMPLE

Write the equation for the reaction showing the lattice enthalpy of sodium oxide.

STEP 1 Work out the formula of sodium oxide:

$$Na_2O$$

STEP 2 See what the gaseous ions must be:

$$Na^+(g) \text{ and } O^{2-}(g)$$

STEP 3 Write the equation for formation of the compound. Remember, there must be only 1 mole of $Na_2O(s)$:

$$2Na^+(g) + O^{2-}(g) \rightarrow Na_2O(s)$$

✔*Quick check 2*

STEP 4 Check the state symbols are correct

i.e. gaseous ions to solid compound.

Remember the charges on ions

Cations	
Sodium	Na^+
Lithium	Li^+
Potassium	K^+
Ammonium	NH_4^+
Magnesium	Mg^{2+}
Calcium	Ca^{2+}
Barium	Ba^{2+}
Silver	Ag^+
Iron(II)	Fe^{2+}
Iron(III)	Fe^{3+}
Lead	Pb^{2+}
Copper(II)	Cu^{2+}

Anions	
Hydroxide	OH^-
Nitrate	NO_3^-
Fluoride	F^-
Chloride	Cl^-
Bromide	Br^-
Iodide	I^-
Sulfate	SO_4^{2-}
Sulfite	SO_3^{2-}
Oxide	O^{2-}
Carbonate	CO_3^{2-}
Sulfide	S^{2-}
Phosphate	PO_4^{3-}

QUICK CHECK QUESTIONS

1 (a) Predict and explain which compound has the more exothermic lattice enthalpy, NaCl or CsCl.
   (b) Predict and explain which compound has the more exothermic lattice enthalpy, NaF or NaBr.

2 Write an equation for the reaction representing the lattice enthalpy of magnesium chloride.

3 Magnesium oxide (MgO) has a much more exothermic lattice enthalpy than sodium chloride (NaCl). Explain why this might be so.

The Born–Haber cycle

Key words

- Born–Haber cycle
- enthalpy change of formation
- Hess' law
- atomisation enthalpy
- ionisation enthalpy
- electron affinity
- lattice enthalpy

Examiner tip

This is a good place to revise general Hess cycle theory. Can you work out which two ways to go round the cycle?

✓ *Quick check 1*

Lattice enthalpies can be calculated from a special type of Hess cycle called a **Born–Haber cycle**. Born–Haber cycles all have the same format. The OCR way of setting them out is shown here. Only one **enthalpy change** occurs in each step of the cycle, and this is shown in bold.

The elements always start at zero energy in their standard states.

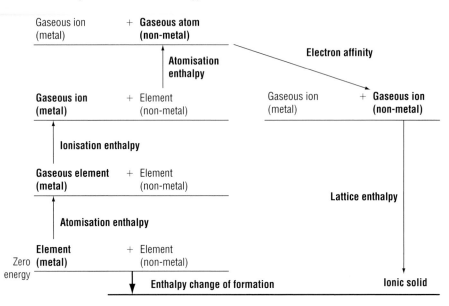

As for all Hess cycles, there are two ways round.

- The first way round simply involves the enthalpy change of formation.
- The second way round involves *atomisation* (to get gaseous atoms) followed by *ionisation* (to get gaseous ions) and finally the *lattice enthalpy*.

So, according to **Hess' law**,

enthalpy change of formation = atomisation enthalpies + ionisation enthalpies + electron affinities + lattice enthalpy

and this can be rearranged to give

lattice enthalpy = enthalpy change of formation – atomisation enthalpies – ionisation enthalpies – electron affinities

You must be aware which energy changes are exothermic and which are endothermic, because exothermic changes go *down* in the Born–Haber cycle and endothermic changes go *up*.

Exothermic	Lattice enthalpy
	Enthalpy change of formation
	Electron affinity (for singly charged anions)
Endothermic	Enthalpy change of atomisation; ionisation energy for metals
	Second electron affinity (for doubly charged anions)

The Born–Haber cycle for NaCl

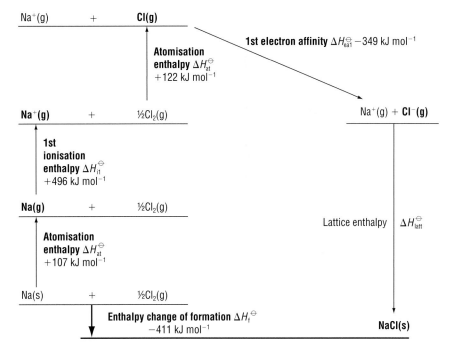

According to Hess' law,

Lattice enthalpy $\Delta H^{\ominus}{}_{\text{latt}} = \Delta H^{\ominus}{}_{\text{f}}$ (sodium chloride)

$-\Delta H^{\ominus}{}_{\text{at}}$ **(sodium)** $-\Delta H^{\ominus}{}_{\text{i1}}$ **(sodium)** $-\Delta H^{\ominus}{}_{\text{at}}$ **(chlorine)** $-\Delta H^{\ominus}{}_{\text{ea1}}$ **(chlorine)**

Substituting,

Lattice enthalpy $= (-411) - (+107) - (+496) - (+122) - (-349) = -787$ kJ mol^{-1}

> **Examiner tip**
>
> The equations representing the changes are:
>
> $\Delta H_{\text{f}}^{\ominus}$(NaCl): Na(s) + ½Cl$_2$(g) → NaCl(s)
>
> ΔH_{at} (Na): Na(s) → Na(g)
>
> $\Delta H_{\text{i1}}^{\ominus}$: Na(g) → Na$^+$(g) + e$^-$
>
> ΔH_{at} (F): ½ F$_2$(g) → F(g)
>
> $\Delta H_{\text{ea1}}^{\ominus}$: F(g) + e$^-$ → F$^-$(g)

> **Examiner tip**
>
> Be careful here with the + and – signs before the numbers. It is best to use brackets as in this example.

  Quick check 2 and 3

Module 2

QUICK CHECK QUESTIONS

1 Construct a labelled Born–Haber cycle for lithium fluoride.

2 Given the data below, calculate the lattice enthalpy for lithium fluoride.

	Enthalpy change	Value/kJ mol^{-1}
I	Standard enthalpy change of formation of lithium fluoride	–616
II	Enthalpy change of atomisation of lithium	+159
III	1st ionisation energy of lithium	+520
IV	Enthalpy change of atomisation of fluorine	+79
V	1st electron affinity of fluorine	–328

3 Give the equations for the enthalpy changes in I to V in question 2.

Things to know about Born–Haber cycles and hydration enthalpies

Key words

- Born–Haber cycle
- enthalpy change of formation
- lattice enthalpy
- ionisation energy
- electron affinity
- exothermic
- endothermic
- enthalpy change of solution
- enthalpy change of hydration

Using the Born–Haber cycle to find the enthalpy change of formation

- The **Born–Haber cycle** can be used to find the **enthalpy change of formation**, $\Delta H^{\ominus}_f$, if all the other terms are known. Just rearrange the equation.
- This can be useful, as it is not always easy to measure $\Delta H^{\ominus}_f$ experimentally.
- As you know, it is not possible to measure the **lattice enthalpy** in the laboratory either, but the lattice enthalpy can be calculated theoretically and then substituted into the equation to find $\Delta H^{\ominus}_f$.

What affects the size of the enthalpy change of formation?

- A large **exothermic** enthalpy change of formation indicates a stable compound.
- The largest exothermic enthalpy change affecting the enthalpy change of formation is the lattice enthalpy.
- The largest **endothermic** enthalpy change is the **ionisation energy**.

As a rough guide, the balance between these two gives the final value for ΔH_f.

Get the ionisation energies right!

- Some compounds contain metal ions with a 2+ charge, such as Mg^{2+}. The Born–Haber cycle in this case involves the *first ionisation enthalpy*, $\Delta H^{\ominus}_{i1}$ ($Mg(g) \rightarrow Mg^+(g) + e^-$), followed by the *second ionisation enthalpy*, $\Delta H^{\ominus}_{i2}$ ($Mg^+(g) \rightarrow Mg^{2+}(g) + e^-$).
- Don't put $Mg(s) \rightarrow Mg^{2+}(g) + 2e^-$! Ionisation enthalpies involve removing *one mole of electrons at a time*.
- Remember that ionisation enthalpies have a positive value for ΔH. They are endothermic.

Examiner tip

This is a good place to revise the definitions of different ionisation energies.

Are electron affinities endothermic or exothermic?

- Some compounds involve a non-metal ion with a 2− charge, such as O^{2-}.
- The Born–Haber cycle in this case involves: the *first **electron affinity***, $O(g) + e^- \rightarrow O^-(g)$, followed by the *second electron affinity*, $O^-(g) + e^- \rightarrow O^{2-}(g)$.
- The first electron affinity is *exothermic* – energy is released when an electron is added.
- The second electron affinity is *endothermic* – energy must be added to get the negatively charged electron to add to the negatively charged ion.

✓ *Quick check 1*

Enthalpy changes of solution and hydration

Definitions

The **enthalpy change of solution** (ΔH_s) is the enthalpy change when 1 mol of a compound is completely dissolved in water under standard conditions.

The **enthalpy change of hydration** (ΔH_{hyd}) is the enthalpy change when 1 mol of isolated gascons ions is dissolved in water to form 1 mol of aqueous ions under standard conditions.

Hydration enthalpies are always exothermic (i.e. negative) because they involve the electrostatic attraction between a charged ion and the dipoles on water molecules (see diagram opposite).

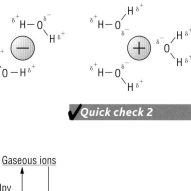

The magnitude of ΔH_{hyd} increases as the charge density on the ion increases, and therefore small ions, such as lithium, have more exothermic hydration enthalpies than larger ions with the same charge, such as potassium. Also, it follows that ions with more than one charge have more exothermic hydration enthalpies. For example, K^+ and Ba^{2+} ions have similar sizes, but the hydration energy of the Ba^{2+} ion is over four times more exothermic.

✓ *Quick check 2*

These rules apply equally to negative ions.

The hydration enthalpy of an ion cannot be determined experimentally and we have to determine it using another Born–Haber cycle, this time involving lattice enthalpy and the enthalpy change of solution. The diagram shows the Born–Haber cycle for an exothermic value of ΔH_s.

```
                              Gaseous ions
                                   ↑
 −Lattice enthalpy              │        │
 (−ΔH_latt)alp                  │        │  Enthalpy change of hydration
 Energy is put in               │        │  of positive and negative ions
 to break the lattice           │        │  ΔH_hyd(+ve ions)+ΔH_hyd(−ve ions)
                    Ions in solid│        │
 Enthalpy change                │        │
 of solution                    │        │
 (ΔH_s)                         ↓        ↓
                              Hydrated ions
```

Therefore, using Hess' law: $\Delta H_{hyd} - \Delta H_{latt} = \Delta H_{solution}$

The cycle above can be used to calculate the ΔH_{hyd} of an ion if all the other quantities are known.

✓ *Quick check 3*

■ WORKED EXAMPLE

For KCl: $\Delta H_{latt} = -711$ kJ mol^{-1}; $\Delta H_{solution} = +22$ kJ mol^{-1}; ΔH_{hyd} for $K^+ = -305$ kJ mol^{-1}.

What is the enthalpy change of hydration of the Cl^- ion?

STEP 1 Use the Born–Haber cycle $\Delta H_{hydration} - \Delta H_{latt} = \Delta H_{solution}$

Also $\Delta H_{hyd} = [\Delta H_{hyd}(K^+) + \Delta H_{hyd}(Cl^-)]$ because there are two ions being hydrated

Therefore $[\Delta H_{hyd}(K^+) + \Delta H_{hyd}(Cl^-)] - \Delta H_{latt} = \Delta H_{solution}$

STEP 2 Rearrange the relationship:

$\Delta H_{hyd}(Cl^-) = \Delta H_{solution} + \Delta H_{latt} - \Delta H_{hyd}(K^+)$

STEP 3 Substitute in the values and complete the calculation – do not forget the units:

$\Delta H_{hyd}(Cl^-) = +22 - 711 - (-305) = -384$ kJ mol^{-1}

Examiner tip

$-\Delta H_{latt}$ is positive here because energy is put in to produce gaseous ions.

✓ *Quick check 4*

QUICK CHECK QUESTIONS

1 Calculate the enthalpy change of formation of calcium chloride using relevant enthalpy change values from the diagram above, plus:
atomisation enthalpy of calcium = +178 kJ mol^{-1}
1st ionisation enthalpy of calcium = +590 kJ mol^{-1}
2nd ionisation enthalpy of calcium = +1145 kJ mol^{-1}
lattice enthalpy of calcium chloride = −2258 kJ mol^{-1}
1st electron affinity of chlorine = −348.8 kJ mol^{-1}
atomisation energy of chlorine = 121.7 kJ mol^{-1}

2 Which ion, Li^+ or Na^+, has the more exothermic hydration enthalpy? Explain your answer.

3 The ions Ca^{2+} and Na^+ have the same ionic radii. Explain which one has the more exothermic enthalpy change of hydration.

4 Calculate the hydration enthalpy of the sodium ion given the following data:
$\Delta H_{hyd}(Cl^-) = -384$ kJ mol^{-1}; $\Delta H_s = +1$ kJ mol^{-1};
$\Delta H_{latt}(NaCl) = -770$ kJ mol^{-1}.

Enthalpy and entropy

You will know many of the concepts used here but **entropy** and **free energy** are new.

- **Enthalpy** change, ΔH, is the heat change at constant pressure.
- **Entropy**, S, is associated with the randomness or disorder of a system.

The entropy of different states of matter

The more freely the particles move in a system, the greater the entropy of the system. The diagram below shows how the entropy changes as the state of matter changes.

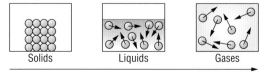

Solids   Liquids   Gases

Increasing freedom of movement
Increasing disorder
increasing entropy

The following examples show how the entropy changes occur:

Change	Example	Entropy change	Explanation
The particles in a solution form a gas	Ammonia gas escaping from solution	Positive	The gas molecules have more freedom than those in solution.
Solid dissolving	Sodium chloride dissolving in water	Positive	There is more disorder in the solution than in the solid.
The number of gas particles decreases	Nitrogen dioxide molecules forming dinitrogen tetroxide $2NO_2(g) \rightleftharpoons N_2O_4(g)$	Negative	There are fewer gas molecules present in the product, therefore less disorder.

The entropy change in a chemical reaction

In a chemical reaction the entropy change (ΔS) can be calculated using the molar entropies for the products and reactants.

$$\Delta S = S_{products} - S_{reactants}$$

The examples below illustrate how this works:

Reaction	$S_{reactant}$ $J\,K^{-1}mol^{-1}$	$S_{product}$ $J\,K^{-1}mol^{-1}$	ΔS $J\,K^{-1}mol^{-1}$	Comment
$CaCO_3(s) \rightleftharpoons CaO(s) + CO_2(g)$	92.9	253.3 (CaO + CO₂)	+160.4	Entropy increases because gas molecules are formed from a solid and therefore the disorder increases.
$2NO_2(g) \rightleftharpoons N_2O_4(g)$	2 × 240.0	304.2	−175.8	Entropy decreases because the number of gas molecules and therefore the disorder decreases.
$NH_3(g) + HCl(g) \rightleftharpoons NH_4Cl(s)$	379.1 (NH₃ + HCl)	94.6	−284.5	Entropy decreases because gas molecules form a solid and therefore there is a decrease in disorder.

Entropy and the second law of thermodynamics

The second law of thermodynamics states that in a spontaneous chemical process *the entropy of the Universe must increase.*

Reaction/change	Exothermic/endothermic	ΔS_{surr} (surrounding)	ΔS_{sys} (system)	Why is the change spontaneous?
$NH_3(g) + HCl(g) \rightleftharpoons NH_4Cl(s)$	Exothermic	Increases	Decreases – gas molecules from a solid	Even though ΔS_{sys} decreases, it is offset by the increase in ΔS_{surr} and $\Delta S_{universe}$ increases.
$NH_4Cl(s) + aq \rightleftharpoons NH_4Cl(aq)$	Endothermic	Decreases	Increases as solid particles become more free to move	The decrease in ΔS_{surr} is offset by the increase in ΔS_{sys} and $\Delta S_{universe}$ increases.

The second law can be put in another form that is easier to calculate. Under standard conditions an equation can be written as follows:

$$\Delta G = \Delta H - T\Delta S$$

ΔG is the *free energy change* of the system. ΔG must be negative in value for the reaction to proceed spontaneously.

To understand how this works we can consider the reaction between ammonia and hydrogen chloride at 298 K.

■ WORKED EXAMPLE

$NH_3(g) + HCl(g) \rightleftharpoons NH_4Cl(s)$

STEP 1 Calculate the enthalpy change for the reaction

$$NH_3(g) + HCl(g) \rightleftharpoons NH_4Cl(s)$$
ΔH_f/kJ mol^{-1} −46.1 −92.3 −314.4

$\Delta H^{\ominus}_r = -314.4 - [-46.1 + (-92.3)] = -176$ kJ mol^{-1} = −176 000 J mol^{-1}

STEP 2 Calculate the entropy change for the reaction

$$NH_3(g) + HCl(g) \rightleftharpoons NH_4Cl(s)$$
ΔS/J K^{-1} mol^{-1} 192.3 186.8 94.6

$\Delta S^{\ominus} = 94.6 - (192.3 + 186.8) = -284.5$ J K^{-1} mol^{-1}

STEP 3 Substitute these values into the free energy equation:

$\Delta G = \Delta H - T\Delta S = -176\,000 - (298 \times -284.5)$ J = −91219 J mol^{-1}
= −91.2 kJ mol^{-1}

The value is negative and therefore the reaction is spontaneous at 298 K.

What would happen if we raised the temperature to 500 K?

$\Delta G = \Delta H - T\Delta S = -176\,000 - (500 \times -164.2) = +6450$ J mol^{-1} = + 6.45 kJ mol^{-1}

ΔG is positive and the reaction will not proceed spontaneously at this temperature!

QUICK CHECK QUESTION

1 For each of the reactions in the table explain the change in entropy and, using the free energy equation, state whether or not the reaction will proceed spontaneously at 298 K.

	Reaction	ΔS J K^{-1}mol^{-1}	ΔH kJ mol^{-1}
(a)	$MgCO_3 \rightleftharpoons MgO + CO_2$	+121	+100
(b)	$C_2H_4(g) + H_2(g) \rightleftharpoons C_2H_6(g)$	−120.6	−137

UNIT 2
Redox reactions and electrode potentials

Key words

- redox reactions
- half-equations
- electrode potentials

You will need to revise reduction, oxidation and **redox reactions** for this section.

Remember that in terms of electron transfer, oxidation is loss of electrons and reduction is gain of electrons. In terms of oxidation numbers, oxidation is an increase in oxidation number and reduction is a decrease.

Redox reactions can be balanced using oxidation numbers. The main point to remember is that the total decrease in oxidation number for one element of the reactants must equal the total increase in oxidation number for another element in the reactants. In the worked example below the oxidation number of the aluminium increases by 3 and that of the iron decreases by 2. To balance these we need 2 aluminium atoms (a total charge of +6) and 3 iron(II) ions (a total charge of −6).

Half-equations and balanced redox equations

Half-equations are used to represent redox equilibria as shown in the table below:

Examiner tip

Remember OILRIG

Oxidation Is Loss of electrons
Reduction Is Gain of electrons

✔ *Quick check 1*

✔ *Quick check 3*

The table below gives examples of these.

System	Example	Half-equation to represent example
A metal in contact with its ions in aqueous solution	Mg and Mg^{2+}	$Mg^{2+} + 2e^- \rightleftharpoons Mg$
A non-metal in contact with its ions in aqueous solution	Cl_2 and Cl^-	$Cl_2 + 2e^- \rightleftharpoons 2Cl^-$
A mixture of ions of the same element in different oxidation states	Fe^{2+} and Fe^{3+}	$Fe^{3+} + e^- \rightleftharpoons Fe^{2+}$

Half-equations can be used to write full balanced equations for redox reactions.

Examiner tip

The convention is to put the electrons on the left-hand side of the half-equation.

■ WORKED EXAMPLE

Write a balanced equation for the reaction between Fe^{2+} ions and aluminium metal to form metallic iron and H^{3+} ions.

STEP 1 Write down the products of the reaction and their formulae.
In this reaction the products are metallic iron (Fe) and aluminium ions (Al^{3+}).

STEP 2 Construct the half-equations involved:

$Fe^{2+} + 2e^- \rightarrow Fe$ **Equation A**
$Al \rightarrow Al^{3+} + 3e^-$ **Equation B**

STEP 3 Multiply equation **A** by **3** and equation **B** by **2**. Then we add the reactants (left-hand sides of both half-equations) together and the products (right-hand sides) together.

$3Fe^{2+} + 6e^- + 2Al \rightarrow 3Fe + 2Al^{3+} + 6e^-$

STEP 4 Remove the electrons from the equation.

$3Fe^{2+} + 2Al \rightarrow 3Fe + 2Al^{3+}$

✔ *Quick check 2*

Redox reactions, electrode potentials and half-cells

- When a metal is in contact with a solution of its ions, an equilibrium is established that results in electrons accumulating on the surface and the metal surface developing a negative charge.

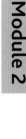

- The more reactive the metal, the greater the tendency to lose electrons, the more negative the charge on the metal surface and hence the more negative the electrode potential of the metal electrode.

- A more reactive metal (e.g. magnesium) will tend to lose electrons more readily than a less reactive metal (e.g. copper).

- The metal in contact with its ions in aqueous solution is called a *half–cell*. Even though the electrons are produced on the surface of the metal, these electrons cannot flow unless the metal is connected to another half-cell. Two half-cells connected together form an *electrochemical cell*.

✔ *Quick check 5*

Electrochemical cells

Any reversible system that contains chemical species undergoing a redox reaction can act as an electrode.

- An electrochemical cell can be constructed from two half-cells. A diagram of an example of an electrochemical cell and the measurement of the voltage produced is shown opposite.

- The voltmeter has a very high resistance. This reduces the current flowing to near zero and the voltage obtained is the e.m.f. of the cell.

- The salt bridge is usually filter paper soaked in saturated potassium nitrate solution. This completes the circuit (by flow of ions) and maintains the ionic balance within each half-cell.

- In all cells the electrons flow from the more negative electrode to the more positive along the wire.

- In this cell the two half-equations are as follows:

$$Mg^{2+} + 2e^- \rightleftharpoons Mg \qquad Cu^{2+} + 2e^- \rightleftharpoons Cu$$

- The half-cell reaction involving Mg^{2+}/Mg has a greater tendency to proceed to the left whilst Cu^{2+}/Cu has a greater tendency to proceed to the right. Therefore electrons flow from the magnesium to the copper through the voltmeter or external circuit.

- The magnesium is the negative electrode and the copper is the positive electrode.

When comparing redox systems other than metal/metal ions a different arrangement has to be used. Two alternative half-cell arrangements are shown here:

✔ *Quick check 3*

Examiner tip

Electrons flow from the negative electrode to the positive electrode.

Aqueous ions (or molecules and ions) with different oxidation states e.g. $Cr^{3+}(aq) + e^- \rightleftharpoons Cr^{2+}(aq)$ $I_2(aq) + 2e^- \rightleftharpoons 2I^-(aq)$	Gaseous elements and their ions e.g. $Cl_2(g) + 2e^- \rightleftharpoons 2Cl^-(aq)$ Chlorine gas is passed into a solution of chloride ions

✔ *Quick check 4*

QUICK CHECK QUESTIONS

1. Define the term reduction in terms of electron transfer and oxidation state.

2. In the reaction $I_2(aq) + 2e^- \rightleftharpoons 2I^-(aq)$, explain why the reverse reaction is an oxidation.

3. Draw the electrochemical cell composed of the Ag/Ag^+ and Cu/Cu^{2+} redox systems and explain the function of the salt bridge.

4. Complete the half-equations below by inserting the missing species. An example is done for you.
   Example: $Mg^{2+} \rightleftharpoons Mg$. The half-equation is completed by adding $2e^-$ to the left-hand side.

   $$Mg^{2+} + 2e^- \rightleftharpoons Mg$$

   (a) $Br_2 + 2e^- \rightleftharpoons$ (b) $\rightleftharpoons Fe^{3+} + e^-$
   (c) $H_2 \rightleftharpoons 2e^- +$ (d) $Mn^{3+} \rightleftharpoons Mn^{2+}$

5. Why is a copper electrode in contact with a solution of copper sulfate described as a half-cell?

Standard electrode potentials

Key words

- standard hydrogen electrode
- standard electrode potential

✓ *Quick check 1*

To compare the electrode potentials of individual half-cells a standard reference electrode has to be used. The one used is the **standard hydrogen electrode** (see opposite).

The properties of this electrode are as follows:

- an inert electrode of platinum black on platinum
- a solution of 1 mol dm⁻³ HCl
- hydrogen gas bubbled into the solution under a pressure of 1 atmosphere (100 kPa)
- the equation for the half-cell is
  $2H^+(aq) + 2e^- \rightleftharpoons H_2(g)$.
- it is assigned a standard electrode potential of 0 V.

The standard electrode potential of a redox system is defined as follows:

- **The standard electrode potential ($E^{\ominus}$)** of a half-cell is the e.m.f. of the half-cell. A standard hydrogen electrode is used as the reference electrode.
- All measurements are at 298 K, 100 kPa, and all solutions are of 1.00 mol dm⁻³ concentration.

✓ *Quick check 2*

A typical arrangement for finding the $E^{\ominus}$ of a redox system is shown below:

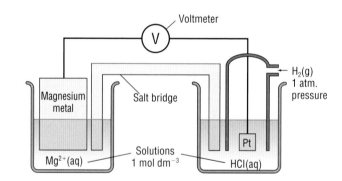

Examiner tip

The $E^{\ominus}$ value does not depend on the number of electrons involved in the reaction.

	Some $E^{\ominus}$ values	
	System	$E^{\ominus}$
Increasing tendency to accept electrons and decreased tendency to release electrons	$Mg^{2+}(aq) + 2e^- \rightleftharpoons Mg(s)$	−2.37 V
	$Zn^{2+}(aq) + 2e^- \rightleftharpoons Zn(s)$	−0.76 V
	$2H^+(aq) + 2e^- \rightleftharpoons H_2(g)$	0.00 V
	$Cu^{2+}(aq) + 2e^- \rightleftharpoons Cu(s)$	+0.34 V
	$Fe^{3+}(aq) + e^- \rightleftharpoons Fe^{2+}(aq)$	+0.77 V
	$Ag^+(aq) + e^- \rightleftharpoons Ag(s)$	+0.80 V
	$Cl_2(g) + 2e^- \rightleftharpoons 2Cl^-(aq)$	+1.36 V

The e.m.f of the cell above is the difference between the $E^{\ominus}$ values for the two half-cells.

$$E_{cell} = 0.00\ V - (-2.37\ V) = 2.37\ V$$

$E^{\ominus}$ shows that if a cell is made up of the Cu/Cu²⁺ and Fe²⁺/Fe³⁺ half-cells, then the copper will form the negative electrode and the electrode for Fe²⁺/Fe³⁺ will form the positive.

✓ *Quick check 3*

Uses of standard electrode potentials

Use I To find the e.m.f. of a cell

■ **WORKED EXAMPLE**

What is the e.m.f. of the cell formed using Cu/Cu^{2+} and Zn/Zn^{2+} half-cells?

STEP 1 Write down the two half-equations involved and their $E^\ominus$ values:

$$Zn^{2+} + 2e^- \rightleftharpoons Zn(s) \qquad E^\ominus = -0.76\,V$$
$$Cu^{2+} + 2e^- \rightleftharpoons Cu(s) \qquad E^\ominus = +0.34\,V$$

STEP 2 To find the e.m.f. of a cell we simply calculate the difference between the $E^\ominus$ values for the half-cells.

$$E_{cell} = +0.34\,V - (-0.76\,V) = 1.10\,V$$

✔*Quick check 4*

Use II To predict the feasibility of a reaction under standard conditions

■ **WORKED EXAMPLE**

Will the reaction below take place?

$$Cu(s) + 2Ag^+(aq) \rightarrow Cu^{2+}(aq) + 2Ag(s)$$

STEP 1 Write down the two half-equations involved and their $E^\ominus$ values:

$$Cu^{2+} + 2e^- \rightleftharpoons Cu(s) \qquad E^\ominus = +0.34\,V$$
$$Ag^+ + e^- \rightleftharpoons Ag(s) \qquad E^\ominus = +0.80\,V$$

STEP 2 Work out what has to happen for the reaction to occur.
The copper has to donate electrons to the silver ions and reduce them to silver.

STEP 3 Work out which way the electrons will flow and therefore the feasibility.

The top half-cell has a more negative $E^\ominus$ value so electrons can be released by the copper, reducing the silver ions. The top reaction proceeds to the left, the bottom one to the right, so the reaction takes place.

✔*Quick check 5*

The effect of non-standard conditions on the e.m.f. of a cell

For example, in a metal/metal ion equilibrium:

$$M^{n+}(aq) + ne^- \rightleftharpoons M(s)$$

Diluting the solution reduces the concentration of the M^{n+} ion and thus disturbs the equilibrium, which shifts to the left to compensate, forming more electrons and therefore making the half-cell electrode potential more negative.

✔*Quick check 6*

QUICK CHECK QUESTIONS

1 Draw a hydrogen electrode and give its main characteristics.

2 Define the term 'standard electrode potential'.

3 What would be the polarity of each half-cell in a cell comprising the half-cells Zn/Zn^{2+} and $Cl_2/2Cl^-$?

4 Calculate the e.m.f. values for the cells made from the following half-cells under standard conditions:

   (a) $Mg^{2+}(aq) + 2e^- \rightleftharpoons Mg(s)$ and $Cl_2(g) + 2e^- \rightleftharpoons 2Cl^-(aq)$
   (b) $Zn(aq) + 2e^- \rightleftharpoons Zn(s)$ and $Fe^{3+}(aq) + e^- \rightleftharpoons Fe^{2+}(aq)$
   (c) $Cl_2(g) + 2e^-(aq) \rightleftharpoons 2Cl^-$ and $Fe^{3+}(aq) + e^- \rightleftharpoons Fe^{2+}(aq)$

5 Is the reaction $2Fe^{2+}(aq) + Zn^{2+}(aq) \rightarrow Fe^{3+}(aq) + Zn(s)$ feasible? Explain your answer.

6 What would happen to the E_{cell} from 4(a) if the pressure of chlorine were increased to above standard pressure?

UNIT 2

Fuel cells, fuel cell vehicles and the hydrogen economy

Key words

- fuel cells

✓ *Quick check 1*

- Modern fuel cells are based on the oxidation of hydrogen or hydrogen-rich fuels to make electrical energy.
- A fuel cell operates like a battery, but as long as the supply of hydrogen and oxygen required for the reactions within the cell is maintained it will not run down.
- A fuel cell consists of an electrolyte sandwiched between two electrodes.
- Oxygen passes over one electrode and hydrogen (or hydrogen-rich fuel) over the other, whilst the electrolyte remains in the cell.

How do fuel cells work?

The equilibria present depend on the fuel and electrolyte used. Three examples are given below.

Type of cell	Electrolyte	Equilibria present and comments
Proton exchange membrane cell	Polymer membrane	$2e^- + 2H^+(aq) \rightleftharpoons H_2(g)$ $E^\ominus = 0.00$ V $2e^- + \frac{1}{2}O_2(g) + 2H^+ \rightleftharpoons H_2O(l)$ $E^\ominus = +1.23$ V At the negative electrode the hydrogen dissociates into H^+ ions and electrons. The ions diffuse through an ion-permeable membrane to the other electrode where they react with O_2 to give water. The overall reaction is: $H_2(g) + \frac{1}{2}O_2(g) \rightarrow H_2O(l)$, i.e. the combustion of hydrogen gas, but instead of producing heat energy, electrical energy is produced. $E_{cell} = E^\ominus = 1.23$ V – 0.00 V = 1.23 V
Alkaline fuel cell	Aqueous alkaline solution (e.g. KOH(aq))	$2e^- + 2H_2O(l) \rightleftharpoons H_2(g) + 2OH^-(aq)$ $E^\ominus = -0.83$ V $2e^- + \frac{1}{2} O_2(g) + H_2O(l) \rightleftharpoons 2OH^-(aq)$ $E^\ominus = +0.40$ V The overall reaction is the same as above and the E_{cell} is again 1.23 V.
Direct methanol fuel cell	Polymer membrane	$6e^- + 6H^+(aq) + CO_2(g) \rightleftharpoons CH_3OH(l) + H_2O(l)$ $6e^- + \frac{3}{2}O_2(g) + 6H^+(aq) \rightleftharpoons 3H_2O(l)$ The overall reaction is $\frac{3}{2}O_2(g) + CH_3OH(l) \rightleftharpoons CO_2(g) + 2H_2O(l)$ This is therefore another combustion reaction. The fuel here is methanol.

✓ *Quick check 2*

A diagram of a fuel cell is shown below.

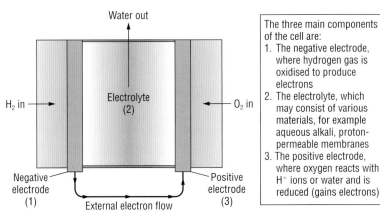

The three main components of the cell are:
1. The negative electrode, where hydrogen gas is oxidised to produce electrons
2. The electrolyte, which may consist of various materials, for example aqueous alkali, proton-permeable membranes
3. The positive electrode, where oxygen reacts with H^+ ions or water and is reduced (gains electrons)

Fuel-cell vehicles (FCVs)

Because of the finite nature of fossil fuels and global warming, vehicles are being developed that are being powered by fuel cells. These fuel cells are powered by:

- hydrogen gas
- hydrogen-rich fuels.

The advantages and disadvantages of FCVs over conventional fossil-fuel cars are outlined below.

Factor to consider	Advantages/positive factors	Disadvantages/negative factors
Safety	Hydrogen leakage can be detected but the detectors are expensive.	Hydrogen is a colourless, odourless gas with small molecules and can therefore leak very easily. It is also flammable.
Pollution and the environment	If hydrogen is used the only product is water. It will also not run out since it can be obtained from water.	The fuel cells themselves are made using toxic chemicals and disposal is therefore a problem.
The storage of the hydrogen	Hydrogen can be either absorbed within certain solid materials or adsorbed onto the surface of materials. In these ways large amounts of hydrogen can be stored in small volumes and at low pressure. The hydrogen can be 'stored' as a hydrogen-rich fuel.	Cylinders of hydrogen are heavy and therefore reduce the efficiency of the vehicle.
Efficiency of the engine	FCVs have an efficiency of about 45% compared with 22% for a diesel vehicle.	The efficiency drops if the hydrogen is stored as high-pressure gas.

Quick check 3 and 4

The hydrogen economy

The hydrogen economy, i.e. the use of hydrogen as a major fuel, is a matter of debate and there are a number of factors to consider:

1. There are problems with the safety, storage and handling of hydrogen (see table above).
2. Hydrogen is an **energy carrier** and not a source of energy. Energy has to be used to extract it from water (by electrolysis) or other sources.
3. It has not yet gained general acceptance as a fuel, and FCVs remain more expensive than their fossil-fuel counterparts.
4. There is currently no infrastructure designed to supply hydrogen as a fuel.

Quick check 5

QUICK CHECK QUESTIONS

1. What are the advantages of a fuel cell over a normal battery?

2. Complete the following half-cell reactions:
   (a) ___ + $2H^+(aq) \rightleftharpoons H_2(g)$
   (b) $2e^- +$ _____ + $2H^+ \rightleftharpoons H_2O(l)$
   (c) ___$e^- + 6H^+(aq) + CO_2(g) \rightleftharpoons CH_3OH(l) +$ _____

3. Methanol is a hydrogen-rich fuel that can be used in fuel cells. Give two disadvantages of using methanol as a fuel instead of hydrogen and one advantage.

4. What are the main ways of storing hydrogen in FCVs?

5. What is the hydrogen economy and what factors need considering if we are to adopt hydrogen as a fuel?

Transition elements – electron configurations

Key words

- transition element
- d-block
- electron configuration

Examiner tip

Remember that *partly filled d-orbital* means d^1 to d^9.

Learn this definition of a transition element.

✔ *Quick check 1*

Examiner tip

[Ar] is $1s^2 \, 2s^2 \, 2p^6 \, 3s^2 \, 3p^6$

Examiner tip

Scandium and zinc are d-block elements but *are excluded from the list of transition elements*. See below for explanation.

✔ *Quick check 2*

Examiner tip

$3d^5$ and $3d^{10}$ are stable because the charge on the electrons is distributed symmetrically around the nucleus.

Transition elements are metals. You will be concerned only with the first row of transition elements, from titanium to copper. They are located in the **d-block** of the Periodic Table because the outermost electrons are in the d sub-shell.

There are many different things to learn about transition elements. Here is a checklist. Make sure you cover them all:

- definition of a transition element
- electron configurations of the elements and their ions
- oxidation states
- catalytic behaviour
- hydroxides
- complex ions
- ligand substitution
- redox reactions and titration calculations.

Definition of a transition element

A transition element has at least one ion with an incomplete d sub-shell.

Electron configuration of the d-block elements

Scandium	Sc	[Ar] $3d^1 \, 4s^2$
Titanium	Ti	[Ar] $3d^2 \, 4s^2$
Vanadium	V	[Ar] $3d^3 \, 4s^2$
Chromium	Cr	[Ar] $3d^5 \, 4s^1$
Manganese	Mn	[Ar] $3d^5 \, 4s^2$
Iron	Fe	[Ar] $3d^6 \, 4s^2$
Cobalt	Co	[Ar] $3d^7 \, 4s^2$
Nickel	Ni	[Ar] $3d^8 \, 4s^2$
Copper	Cu	[Ar] $3d^{10} \, 4s^1$
Zinc	Zn	[Ar] $3d^{10} \, 4s^2$

The 3d sub-shell is filled after the 4s sub-shell, so most of the elements have a full 4s sub-shell, $4s^2$. *But note that chromium and copper have $4s^1$ electron configurations, not $4s^2$.* This is to allow either a half-filled or a filled d sub-shell to be made – Cr has $3d^5 \, 4s^1$ and Cu has $3d^{10} \, 4s^1$. A half-filled or completely filled sub-shell is more stable, so it makes sense in energy terms for chromium and copper to have these electron configurations.

Zinc and scandium are not included in lists of transition metals although they are in the first row of the d-block. This is because:

- Zinc forms one ion, Zn^{2+}, with an electron configuration of $1s^2 \, 2s^2 \, 2p^6 \, 3s^2 \, 3p^6 \, 3d^{10}$. This means that its only ion has a full, not a partially full, d sub-shell – so zinc is not a transition element.

- Scandium forms one ion, Sc^{3+}, with an electron configuration of $1s^2\ 2s^2\ 2p^6\ 3s^2\ 3p^6$. This is the main ion, and it has only empty d orbitals, and so scandium is excluded from the list of transition metals.

- We commonly say that scandium and zinc are not transition elements, but they are d-block elements.

Electron configurations of the ions

The transition elements form positive ions, as they are metals. This means that when an ion is formed, electrons are removed from the atom.

You must remember that the 4s electrons are removed first.

> ### ■ WORKED EXAMPLE 1
>
> Give the electron configuration of Cu^{2+}.
>
> **STEP 1** Write down the electron configuration of the atom, Cu.
>
> **[Ar] 3d¹⁰ 4s¹**
>
> **STEP 2** The ion has a 2+ charge so two electrons are removed. One is taken from the 4s sub-shell, the other from the 3d sub-shell.
>
> So the electron configuration of Cu^{2+} is: $1s^2\ 2s^2\ 2p^6\ 3s^2\ 3p^6\ 3d^9$.

> ### ■ WORKED EXAMPLE 2
>
> Give the electron configuration of Fe^{3+}.
>
> **STEP 1** write down the electron configuration of the atom, Fe.
>
> **[Ar] 3d⁶ 4s²**
>
> **STEP 2** The ion has a 3+ charge so three electrons are removed. Two are taken from the 4s sub-shell, the other from the 3d sub-shell.
>
> So the electron configuration of Fe^{3+} is: $1s^2\ 2s^2\ 2p^6\ 3s^2\ 3p^6\ 3d^5$.

✓ *Quick check 4*

Examiner tip

If a d-block element has a white compound, in that oxidation state it will probably have a completely full or empty d sub-shell. *Colour* is associated with *partly filled d-orbitals*.

Module 3

✓ *Quick check 3 and 4*

QUICK CHECK QUESTIONS

1 Explain what is meant by the term *transition element*.

2 Give the electron configurations of manganese, Mn, and chromium, Cr.

3 Give the electron configurations of Mn^{2+} and Cr^{3+}.

4 Explain why you would not expect compounds of Cu^+ ions to be coloured whilst those of Cu^{2+} are different colours.

Transition elements – oxidation states, catalytic behaviour and the hydroxides

Key words

- oxidation state
- catalysts

Examiner tip

Learn these common oxidation states: for Fe, +2, +3, for Cu, +1, +2.

✔*Quick check 1*

Examiner tip

Fe and V_2O_5 are heterogeneous catalysts that lower the activation energy of the reaction by adsorbing the gas molecules onto their surface using their d-orbitals.

✔*Quick check 2*

Oxidation states

One characteristic of the transition elements is that they form compounds and ions with different **oxidation states**. In forming ions the two electrons in the 4s orbital are lost first. The 3d and 4s energy levels are close in energy and therefore the 3d electrons can also be lost in forming ions.

You have to learn the different oxidation states of iron and copper.

- **Iron** can have +2, +3, +4, +5 and +6 oxidation states. The +2 and +3 oxidation states are the most common.

- **Copper** can have +1, +2 and +3 oxidation states. The +1 and +2 oxidation states are the most common.

Catalytic behaviour

Transition metals and their compounds are very good **catalysts**. There are two reasons for this:

- They can have different oxidation states, so they can gain and lose electrons in moving between these oxidation states, thus facilitating and speeding up redox reactions.

- They provide sites at which reactions can take place, because they bond to a wide range of ions and molecules in solution and as solids.

Examples of industrial catalysts are:

finely divided **Fe** or **Fe_2O_3** in the production of ammonia

$$N_2(g) + 3H_2(g) \rightleftharpoons 2NH_3(g)$$

solid **V_2O_5** in the production of sulfur trioxide, used to make sulfuric acid

$$2SO_2(g) + O_2 \rightleftharpoons 2SO_3(g)$$

finely divided **Ni** in the hydrogenation of alkenes

$$CH_2=CH_2(g) + H_2 \rightarrow CH_3CH_3(g)$$

Transition metal hydroxides

Transition metal ions react with the hydroxide ion in aqueous solution to give a solid.

OH⁻ (aq) + transition metal ion(aq) → metal hydroxide(s)

Example:

The reaction between aqueous copper sulfate and aqueous sodium hydroxide is

$CuSO_4(aq) + 2NaOH(aq) \rightarrow Cu(OH)_2(s) + Na_2SO_4(aq)$

The colour of the metal hydroxide can be used to identify the metal.

You need to know the colours of copper (II) hydroxide, iron(II) and iron(III) hydroxides, and cobalt(II) hydroxide. The list below shows the equations for the reactions and the colours of the aqueous solutions and precipitates obtained.

Cu^{2+} (aq) blue **+ $2OH^-$(aq)** → **$Cu(OH)_2(s)$** blue

Fe^{2+}(aq) pale green **+ $2OH^-$(aq)** → **$Fe(OH)_2(s)$** green

Fe^{3+}(aq) yellow/orange **+ $3OH^-$(aq)** → **$Fe(OH)_3(s)$** red-brown/rust

Co^{2+}(aq) pink **+ $2OH^-$(aq)** → **$Co(OH)_2(s)$** blue-green and then pink

Although these reactions are the only ones you must know, make sure you can predict the reaction between any transition metal ion and aqueous sodium hydroxide.

> ### Examiner tip
> The NaOH(aq) supplies the OH⁻ ions.

Module 3

■ WORKED EXAMPLE

Predict the formula of the precipitate formed when aqueous sodium hydroxide is added to an aqueous solution of $CrCl_3$(aq).

STEP 1 Work out the formula of the transition metal ion:

In $CrCl_3$ the chromium ion must be Cr^{3+}.

STEP 2 The charge on the metal ion is balanced by the number of hydroxide ions in the precipitate:

The charge is 3+ so three OH⁻ ions are needed.

STEP 3 Write the ionic equation:

$Cr^{3+}(aq) + 3OH^-(aq) \rightarrow Cr(OH)_3(s)$

 ✓*Quick check 3*

QUICK CHECK QUESTIONS

1 Explain what is meant by the term *transition element*.

2 Suggest why *finely divided* iron is used as a catalyst in the production of ammonia.

3 **(a)** A transition metal ion in aqueous solution (X) was added to aqueous sodium hydroxide. A brown precipitate appeared. Identify the transition metal ion and write an ionic equation for the reaction.

(b) (i) What is the colour of the aqueous solution of the transition metal ion, X?

(ii) If sulfur dioxide gas is passed through the aqueous solution of X, the solution changes in colour to pale green. What is the ion formed in this reaction?

(iii) What would you see if sodium hydroxide was added to this green solution?

UNIT 2 Complex ions

Module 3

Key words

- complex ions
- ligands
- coordination numbers
- stereoisomerism

✓ *Quick check 1*

Examiner tip

Octahedral and tetrahedral are the commonest shapes but square planar complexes are common for platinum and nickel.

A **complex ion** is a transition metal atom or ion + ligands.

A **ligand** is a molecule or ion which *donates a pair of electrons* to the transition metal ion in a complex to form a dative covalent (or coordinate) bond.

An example of a complex ion is $[Cu(H_2O)_6]^{2+}$.

This formula is used to show that Cu^{2+} is surrounded by six H_2O molecules. The H_2O molecules are the ligands. Square brackets go round the whole complex and the total charge of the complex ion goes outside these square brackets. The ligands are shown in normal brackets, with the number of ligands at the end. The number of coordinate bonds round the central ion is called the **coordination number.**

Shapes of complex ions

Complex ions occur in various shapes. The three most common shapes are shown in the table below:

Octahedral Coordination number 6 Bond angle 90°	Tetrahedral Coordination number 4 Bond angle 109.5°	Square planar Coordination number 4 Bond angle 90°
e.g. $[Cu(H_2O)_6]^{2+}$	e.g. $CuCl_4^{2-}$	e.g. $[Ni(NH_3)_2Cl_2]$

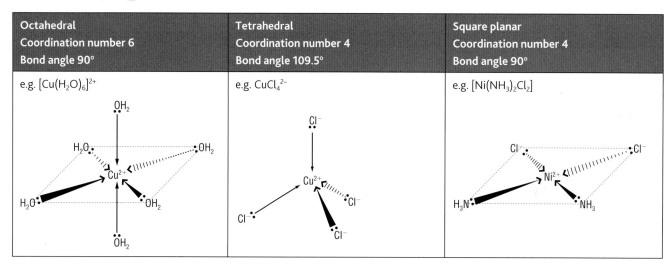

✓ *Quick check 2*

Examiner tip

When you draw a complex ion, make sure you show clearly which atom in the ligand forms the coordinate bond.

Types of ligand

- Ligands can be neutral or anionic (negatively charged).
- Ligands can donate *one* pair of electrons (*mono*dentate), *two* pairs of electrons (*bi*dentate) or *several* pairs of electrons (*multi*dentate).

Here are the most common ligands you will come across.

Examiner tip

EDTA is commonly used in shampoos and other cleaning agents.

Water	H_2O	monodentate
Ammonia	NH_3	monodentate
Chloride ion	Cl^-	monodentate
Hydroxide ion	OH^-	monodentate
Cyanide ion	$CN-$	monodentate
Ethane-1,2-diamine (abbreviated to en)	$H_2NCH_2CH_2NH_2$	bidentate
EDTA^{4-}		multidentate (hexadentate)

✓ *Quick check 3*

Stereoisomerism in transition metal complexes

There are two types of stereoisomerism and you will have come across both types when you studied alkenes at AS level (*cis–trans* as a special case of *E/Z* isomerism) and stereoisomerism (optical isomerism) at A2.

Cis–trans isomerism

Cis–trans isomerism in transition metals is different from that found in the alkenes because there is no double bond present, just the arrangement of ligands around a central transition metal ion. The *cis* isomer has two identical groups lying on one side of the metal ion (with a bond angle of 90° between them), whilst the *trans* isomer has two identical groups lying on opposite sides of the ion (180° apart).

The two examples below illustrate how *cis–trans* isomerism can occur for square planar and octahedral complexes.

Square planar complex – $[Ni(NH_3)_2Cl_2]$	Octahedral complex
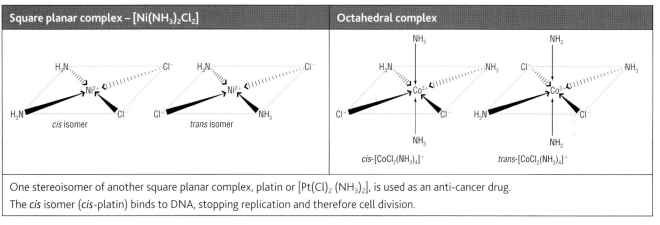	

One stereoisomer of another square planar complex, platin or $[Pt(Cl)_2(NH_3)_2]$, is used as an anti-cancer drug. The *cis* isomer (*cis*-platin) binds to DNA, stopping replication and therefore cell division.

Optical isomerism

✔ *Quick check 4*

In optical isomerism there are two isomers that are non-superimposable mirror images of each other. An example of this is the $[Ni(H_2NCH_2CH_2NH_2)_3]^{2+}$ complex.

The $H_2NCH_2CH_2NH_2$ ligand is a bidentate ligand and can be represented as shown opposite when drawing complexes.

The two isomers of $[Ni(H_2NCH_2CH_2NH_2)_3]^{2+}$ are shown below.

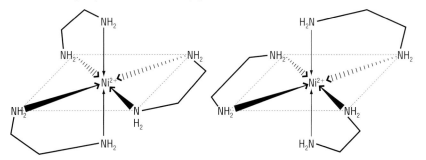

✔ *Quick check 5*

QUICK CHECK QUESTIONS

1   Using $[Cu(H_2O)_6]^{2+}$ as an example, explain the terms ligand, complex, octahedral and coordination number.

2   Draw the following complexes:
    (a) $[Fe(H_2O)_6]^{2+}$
    (b) $CoCl_4^{2-}$.

3   Using $H_2NCH_2CH_2NH_2$ as an example, explain the term bidentate ligand.

4   Draw the two isomers of $[Pt(NH_3)_2Cl_2]$.

5   Draw the two optical isomers of $[CoCl_2(H_2NCH_2CH_2NH_2)_2]$.

Ligand substitution

Key words

- ligand substitution
- stability constant

Some ligands combine more strongly with transition metal ions than others. A ligand that binds strongly can displace a ligand that binds more weakly. This is called **ligand substitution**. You can see ligand substitution in experiments because *different ligands change the colour of the solution as a different complex is formed.*

There are certain ligand substitution reactions you must know, along with the colour changes accompanying them. These are illustrated below. The diagrams show the structures of the complexes and the colour changes occurring.

Example 1 Ammonia solution is added to a solution of aqueous copper(II) ions (e.g. aqueous copper(II) sulfate solution):

Examiner tip

Both these complexes have six ligands, so they are octahedral in shape.

✔ *Quick check 1*

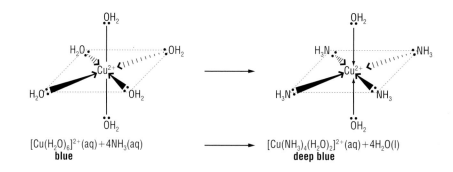

$$[Cu(H_2O)_6]^{2+}(aq) + 4NH_3(aq) \longrightarrow [Cu(NH_3)_4(H_2O)_2]^{2+}(aq) + 4H_2O(l)$$
blue → **deep blue**

Example 2 Concentrated hydrochloric acid (Cl⁻) is added to aqueous copper(II) ions:

✔ *Quick check 2*

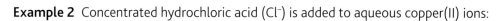

$$[Cu(H_2O)_6]^{2+} + 4Cl^-(aq) \longrightarrow \text{Green due to a mixture of the two complexes} \longrightarrow CuCl_4^{2-}(aq) + 4H_2O(l)$$
blue → → **yellow**

Example 3 Water is added to a solution of $CoCl_4^{2-}$ ions:

Examiner tip

This is used as a test for water. The $CoCl_4^{2-}$ ion is on cobalt chloride paper. When water is added it turns from blue to pink.

Examiner tip

To get the overall charge on the complex $CoCl_4^{2-}$ ion you add together the charge on the metal ion and the charges on the ligands.

Overall charge = charge on metal ion + charges on ligands.

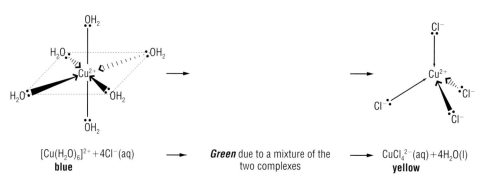

$$CoCl_4^{2-}(aq) + 6H_2O(l) \longrightarrow [Co(H_2O)_6]^{2+} + 4Cl^-(aq)$$
blue → **pink**

Haemoglobin

In haemoglobin a central Fe^{2+} ion is the ion in an octahedral complex comprising four dative covalent bonds from the nitrogens in a porphyrin ring, and one from a nitrogen on one of the amino acids on the globin molecule, which is the protein part of the haemoglobin molecule. *The sixth ligand in the complex is oxygen.*

Carbon monoxide is toxic because it binds to the central Fe^{2+} ion more strongly than oxygen and replaces it in the complex. This means that oxygen cannot be carried around the body, causing asphyxiation.

The reaction is $HbO_2(aq) + CO(g) \rightleftharpoons HbCO(aq) + O_2(g)$

 *Quick check 3*

Stability constants and ligand substitution

When a ligand is added to a solution of the complex formed between water and a transition metal ion, an equilibrium is established as ligand substitution takes place. You have to be able to write the equilibrium constant for this equilibrium, called the **stability constant.**

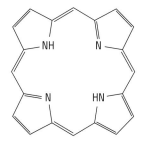

Module 3

■ WORKED EXAMPLE

How do we write the stability constant when Cl^- substitutes for water molecules?

STEP 1 *Write the equation for the reaction.* This always concerns the complex containing water as a reactant:

$$[Cu(H_2O)_6]^{2+} + 4Cl^-(aq) \rightleftharpoons CuCl_4^{2-}(aq) + 6H_2O(l)$$

STEP 2 Write the equilibrium constant for the reaction.

$$K_c = \frac{[CuCl_4^{2-}][H_2O]^6}{[[Cu(H_2O)_6]^{2+}][Cl^-]^4}$$

 *Quick check 1*

STEP 3 *Simplify this by omitting the concentration of water as a factor.* This is because the concentration is a constant. The new equilibrium constant is the stability constant.

$$K_{stab} = \frac{[CuCl_4^{2-}]}{[[Cu(H_2O)_6]^{2+}][Cl^-]^4}$$

STEP 4 Add the units to your expression. In this case we have:

$$units = \frac{mol\ dm^{-3}}{mol\ dm^{-3} \times (mol\ dm^{-3})^4} = \frac{1}{(mol\ dm^{-3})^4} = dm^{12}\ mol^{-4}$$

The magnitude of the stability constant is a measure of how strongly the ligand binds to the central metal ion. The greater K_{stab} the stronger the binding.

The strength of binding of some ligands to Cu^{2+} follows the following trend:

$$EDTA^{4-} > NH_3 > Cl^- > H_2O$$

QUICK CHECK QUESTIONS

1 Using the reaction
$[Cu(H_2O)_6]^{2+}(aq) + 4NH_3(aq) \rightleftharpoons$
$\qquad [Cu(NH_3)_4(H_2O)_2]^{2+}(aq) + 4H_2O(l)$:
(a) Explain the term 'ligand substitution'.
(b) Write the expression for the stability constant.

2 Give the colour changes occurring when Cl^- ions are added to a solution of $[Cu(H_2O)_6]^{2+}(aq)$ ions.

3 Explain why carbon monoxide poisoning is caused by ligand substitution.

Redox reactions and titration calculations

Key words

- redox reactions
- titrations

You have met redox reactions before in the AS course (AS Revision Guide page 16) and on page 72 of this book. Now you will study some redox reactions involving transition metal ions. There are many of these, because transition elements have *several different oxidation states*.

Often *half-equations* and *ionic equations* are used to show redox behaviour.

Redox behaviour in iron

$$Fe^{3+}(aq) + e^- \rightleftharpoons Fe^{2+}(aq)$$
Yellow   Pale green

Species	Oxidation number of Fe
Fe^{2+}	+2
Fe^{3+}	+3

Examiner tip

Use the terms oxidising, reducing, oxidising agent and reducing agent carefully – it's easy to get muddled. The best way of working out what is oxidised and what is reduced is to use *oxidation numbers*.

Iron can change from:

oxidation state +2 to +3 (oxidation) if an oxidising agent is added to it.

oxidation state +3 to +2 (reduction) if a reducing agent is added to it.

- Fe^{3+} is itself an oxidising agent, because it can oxidise other species by gaining an electron.
- Fe^{2+} is itself a reducing agent, because it can reduce other species by donating an electron.

Example:

$$Fe^{3+}(aq) + 2I^-(aq) \rightarrow Fe^{2+}(aq) + I_2(aq)$$

✓ *Quick check 1*

You can tell by the colour change that this reaction has happened. Fe^{3+} is yellow-orange, Fe^{2+} is pale green and I_2 is brown. The brown colour of the I_2 masks the pale green of Fe^{2+}, so the colour change is pale orange to brown.

Redox behaviour in manganese

$$MnO_4^-(aq) + 8H^+(aq) + 5e^- \rightarrow Mn^{2+}(aq) + 4H_2O(l)$$
manganate(VII) ion   manganese(II) ion
purple   very pale pink

Examiner tip

The state symbols have been left out of these equations to make them clearer.

This reaction takes place in acid solution, as you can see by the presence of $H^+(aq)$ ions in the equation.

Manganese is reduced in this reaction, so the manganate(VII) ion is an oxidising agent.

Species	Oxidation number of Mn
MnO_4^-	+7
Mn^{2+}	+2

Example:

$$MnO_4^- + 5Fe^{2+} + 8H^+ \rightarrow Mn^{2+} + 5Fe^{3+} + 4H_2O$$

The colour change here is from purple (MnO_4^-) to yellow (Fe^{3+}) – the very pale pink of the Mn^{2+} does not show.

Redox behaviour in chromium

$$Cr_2O_7^{2-}(aq) + 14H^+(aq) + 6e^- \rightarrow 2Cr^{3+} + 7H_2O$$
dichromate(VI) ion   chromium(III) ion
orange   green

Examiner tip

Notice that all these reactions take place in acid solution.

This reaction is frequently used in organic chemistry, where the orange → green colour change tells you the organic substance has been oxidised (see page 9 and AS Revision Guide page 62).

Species	Oxidation number of Cr
$Cr_2O_7^{2-}$	+6
Cr^{3+}	+3

Titration calculations

You must revise *titration calculations* because they will be tested!

$KMnO_4$ titrations

Make sure you are familiar with the redox titration between MnO_4^- and Fe^{2+} in aqueous acid solution.

- The equation for this reaction is:

$$MnO_4^-(aq) + 5Fe^{2+}(aq) + 8H^+(aq) \rightarrow Mn^{2+}(aq) + 5Fe^{3+}(aq) + 4H_2O(l)$$

- The *purple* aqueous MnO_4^- is added from the burette to the aqueous Fe^{2+}. It immediately goes *colourless* as it reacts with the Fe^{2+} (the very pale pink of $Mn^{2+}(aq)$ and the pale yellow of dilute $Fe^{3+}(aq)$ do not show at these low concentrations).
- The end point is when all the Fe^{2+} has reacted and a *permanent pink colour* can be seen.
- Remember the acid! It takes part in the reaction, so without acid the reaction will not happen.

Thiosulfate titrations

Sodium thiosulfate (containing the $S_2O_3^{2-}$ ion) is a useful reagent because it can be titrated against iodine. It is particularly useful in linked reactions. For example, it can used to estimate the concentration of Cu^{2+} ions in a solution.

Firstly, excess iodide ions are added to the solution of copper(II) ions:

$$2Cu^{2+}(aq) + 4I^-(aq) \rightarrow 2CuI + I_2$$

The iodine liberated is then titrated against standard sodium thiosulfate solution:

$$I_2 + 2S_2O_3^{2-} \rightarrow S_4O_6^{2-} + 2I^-$$

From these two equations it can be seen that:

$$2Cu^{2+} \equiv I_2 \equiv 2S_2O_3^{2-}$$

Therefore $2Cu^{2+} \equiv 2S_2O_3^{2-}$

Therefore the number of thiosulfate ions = the number of copper(II) ions.

We can then use our equations for moles in solution to find the concentration of the copper.

Examiner tip

See page 15 of Revise AS Chemistry for OCR.

Examiner tip

Remember the colour change at the end point is *colourless →* *pink*.

✔*Quick check 2*

Examiner tip

Starch is used as the indicator in this reaction and the end point is *blue-black →* *colourless*.

✔*Quick check 3*

Examiner tip

Use $n = C \times V$ in your calculation.

Module 3

QUICK CHECK QUESTIONS

1 Hydrogen peroxide, H_2O_2, is an oxidising agent.

$$H_2O_2(aq) + 2H^+(aq) + 2e^- \rightarrow 2H_2O(l)$$

Construct the equation for the reaction between hydrogen peroxide and Fe^{2+} ions in aqueous solution.

2 A student weighed out 5.56 g of $FeSO_4 \cdot 7H_2O$, dissolved it in dilute sulfuric acid and made up the solution to 250 cm^3 in a volumetric flask using distilled water. She then titrated 25 cm^3 samples of this solution against potassium manganate(VII) solution. The iron(II) sulfate solution required 20.0 cm^3 for complete reaction.

(a) Calculate the concentration of the Fe^{2+} ion in the iron(II) sulfate solution.

(b) Calculate the concentration of the potassium manganate solution.

3 Brass is an alloy of copper and zinc. To find the percentage of copper in a sample of brass, a student dissolved 2.20 g of the brass in concentrated nitric acid and made the resulting solution up to 250 cm^3 in a volumetric flask. He then took 25.0 cm^3 samples of this solution, added excess potassium iodide solution and then titrated the iodine liberated against a standard, 0.100 mol dm^{-3} solution of sodium thiosulfate. Starch was used as the indicator. The solution required 22.5 cm^3 of sodium thiosulfate for complete reaction.

(a) Give the equations for the reaction of Cu^{2+} ions with I^- ions and the reaction of iodine with thiosulfate ions.

(b) Calculate the concentration of copper ions in the solution and the percentage composition of the sample of brass. (NOTE: zinc does not react with I^- ions.)

UNIT 2

End-of-unit questions

1 Copper readily forms complexes with water, ammonia and chloride ions, and pairs of these ions can coexist in different equilibria.

   (a) Concentrated hydrochloric acid is added to a solution containing $[Cu(H_2O)_6]^{2+}(aq)$ until there is a high concentration of Cl^-.
   $$[Cu(H_2O)_6]^{2+}(aq) + 4Cl^-(aq) \rightleftharpoons [CuCl_4]^{2-}(aq) + 6H_2O(l)$$
   pale blue   yellow

      (i)   Using one or both of the two complexes given above, explain what is meant by the following terms:
            I   coordinate bonding   [2]
            II   ligand   [2]
            III complex   [2]
      (ii) Draw the two complexes described in the equilibrium.   [4]
   (b) (i)   State le Chatelier's principle.   [2]
      (ii) For the equilibrium shown, use le Chatelier's principle to deduce what is observed as the acid is added.   [3]
   (c) The equilibrium constant, K_c, for the equilibrium in (a) may be written as:

      $$K_c = \frac{[[CuCl_4]^{2-}][H_2O]^6}{[[Cu(H_2O)_6]^{2+}][Cl^-]^4}$$

      It is possible in aqueous solution to simplify this expression to:

      $$K'_c = \frac{[[CuCl_4]^{2-}]}{[[Cu(H_2O)_6]^{2+}][Cl^-]^4}$$

      The numerical value of K'_c for this equilibrium, at 25 °C, is 4.17×10^5.
      (i)   What are the units of K'_c?   [1]
      (ii) An equilibrium mixture, at 25 °C, contained 1.17×10^{-5} mol dm^{-3} $[Cu(H_2O)_6]^{2+}(aq)$ and 0.800 mol dm^{-3} Cl^-.
            Calculate the equilibrium concentration of $[CuCl_4]^{2-}(aq)$.   [2]
      (iii) Suggest why, in aqueous solution, it is possible to simplify K_c to K'_c.   [2]
                                                                    [TOTAL 20 marks]

2 (a) (i)   With the aid of examples, explain the meaning of the terms *strong acid* and *weak acid*.   [2]
      (ii) Using the ethanoic acid/sodium ethanoate mixture as an example, explain how a buffer solution works.   [5]
   (b) Assuming the temperature to be 25 °C, what is the pH of:
      (i)   0.05 mol dm^{-3} hydrochloric acid   [1]
      (ii) 0.01 mol dm^{-3} sodium hydroxide?   [2]
   (c) Benzoic acid, C_6H_5COOH, is a weak acid with an acid dissociation constant, K_a, of 6.3×10^{-5} mol dm^{-3} at 25 °C.
      (i)   Calculate the pH of 0.020 mol dm^{-3} benzoic acid at this temperature.   [3]
      (ii) Draw a sketch graph of the change in pH which occurs when 0.020 mol dm^{-3} potassium hydroxide is added to 25 cm^3 of 0.020 mol dm^{-3} benzoic acid until in excess.   [3]
      (iii) The pK_a values of some indicators are:
            thymol blue   1.7
            congo red   4.0
            thymolphthalein   9.7
            Which of these indicators would be most suitable for determining the end point of the titration between the benzoic acid and potassium hydroxide in (ii)? Explain your answer   [2]
                                                                    [TOTAL 18 marks]

3 Methanol, CH_3OH, is used as an alternative fuel to petrol in racing cars. Although methanol is less volatile than petrol, its combustion in these engines is more complete.

(a) Enthalpy changes can be determined indirectly using standard enthalpy changes of formation.

(i) Write the balanced equation for the combustion of methanol. [1]

(ii) Calculate the standard enthalpy change of combustion of methanol using the following data.

Compound	ΔH_f/kJ mol^{-1}
$CH_3OH(l)$	−239
$CO_2(g)$	−394
$H_2O(l)$	−286

[3]

(iii) The molar entropies of the reactants and products involved in the reaction are given below:

	ΔS J K mol^{-1}
$CH_3OH(l)$	127
$CO_2(g)$	214
$H_2O(l)$	70
$O_2(g)$	~~127~~ 103

Using these values calculate the standard entropy change for the reaction. [2]

(b) Calculate the molar free energy change for the reaction at 298 K and hence explain whether or not it is a spontaneous reaction at this temperature. [4]

(c) At 298 K the combustion of methanol requires, for example, a lighted spill or a spark from an ignition coil. In the light of your answer to part **(c)**, explain this fact. [2]

(d) One of the consequences of the heat generated by a car engine is the formation of nitrogen oxide, NO.

When the nitrogen monoxide leaves the car engine it is oxidised to a gas in which the oxidation number of the nitrogen is +4.

(i) Identify the product and write the balanced symbol equation for the reaction. [3]

(ii) Explain why it is a redox reaction. [2]

In an experiment to investigate the effects of changing concentrations on the rate of reaction, the following results were obtained.

Experiment number	Initial concentration O_2/10^{-2} mol dm^{-3}	Initial concentration NO/10^{-2} mol dm^{-3}	Initial rate of disappearance of NO/10^{-4} mol dm^{-3}s^{-1}
1	1.0	1.0	0.7
2	1.0	2.0	2.8
3	1.0	3.0	6.3
4	2.0	2.0	5.6
5	3.0	3.0	18.9

        (iii) Deduce the order of reaction with respect to O_2 and NO, explaining your reasoning. [4]

        (iv) (i) Write the rate equation for this reaction.

           (ii) Calculate the value of the rate constant k and state its units. [3]

           [TOTAL 24 marks]

4 (a) Standard electrode potentials can be used to measure the relative tendencies of redox systems to gain or lose electrons.

    (i) Define the term standard electrode potential. [2]

    (ii) The standard electrode potentials for two half-cell reactions are given below:

$$E^{\ominus}$$

$$e^- + Fe^{3+} \rightleftharpoons Fe^{2+} \quad +0.77 \text{ V}$$
$$e^- + Cu^{2+} \rightleftharpoons Cu^+ \quad +0.15 \text{ V}$$

Using these values, explain whether or not the reaction between $Cu^+(aq)$ and $Fe^{3+}(aq)$ ions will take place as shown in the following equation:

$$Cu^+(aq) + Fe^{3+}(aq) \rightarrow Cu^{2+}(aq) + Fe^{2+}(aq)$$ [3]

(b) Brass is an alloy of copper and zinc. The composition of the brass can be varied in order to change the properties of the alloy.

In order to find the percentage of copper in a sample of brass a student weighed out 2.26 g of the sample.

She then reacted this with concentrated nitric acid to produce a solution containing Zn^{2+} and Cu^{2+} ions.

This solution was then made up to 250 cm^3 in a volumetric flask.

25 cm^3 samples of the solution were then added to a conical flask, and excess potassium iodide solution was added to give a precipitate of copper(I) iodide (CuI) and iodine solution.

The equation for the reaction is:

$$2Cu^{2+}(aq) + 4I^-(aq) \rightarrow 2CuI(s) + I_2(aq)$$

The iodine liberated was then titrated against 0.100 mol dm^{-3} sodium thiosulfate solution using starch as the indicator. The volume of thiosulfate solution required for complete reaction was 22.1 cm^3. The equation for this reaction is:

$$I_2(aq) + 2S_2O_3^{2-} \rightarrow 2I^-(aq) + S_4O_6^{2-}(aq)$$

The equation for the reaction between the nitric acid and the copper in the brass is as follows:

$$Cu(s) + 2NO_3^-(aq) + 4H^+(aq) \rightarrow Cu^{2+}(aq) + 2H_2O(l) + 2NO_2(g)$$

    (i) Give *and explain* two possible observations and explain why the reaction is a redox reaction. [6]

    (ii) Calculate the concentration of the copper in the solution in the conical flask. [3]

    (iii) Calculate the mass of copper present in the sample. [3]

    (iv) Calculate the percentage composition of the brass. [2]

        [TOTAL 19 marks]

5 (a) Define the term lattice enthalpy. [2]

(b) The values for the lattice enthalpies of various compounds are shown below:

Compound	ΔH_{latt}/kJ mol^{-1}
LiCl	–849
NaCl	–781
KCl	–710

    (i) Explain the changes in lattice enthalpy for the chlorides of Group I as the Group is descended. [3]

    (ii) Explain why the hydration enthalpy of all three alkali metal ions is exothermic. [3]

        [TOTAL 8 marks]

Answers to quick check questions

Unit 1   Rings, polymers and analysis

Module 1 – Rings, acids and amines

Arenes

1 (a) Yes
   (b) No
   (c) Yes
   (d) No
2 (a) Carbon–carbon bond lengths are equal and are intermediate between C–C and C=C.
      Benzene undergoes substitution reactions rather than addition.
      Benzene is much more stable than would be expected if it had the Kekulé structure.
   (b) The p-orbitals overlap above and below the plane of the benzene ring to form one continuous π-electron system.

3 (a)   (b)   (c)   (d)   (e)

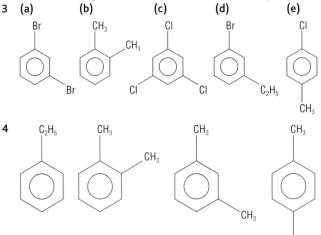

5 (a) chlorobenzene
   (b) 1, 4-dimethylbenzene
   (c) propylbenzene

Electrophilic substitution

1 See text.
2 $C_6H_6 + HNO_3 \rightarrow C_6H_5NO_2 + H_2O$
   For mechanism see text.
3 (a) A = C_6H_5Cl; B = HCl; aluminium chloride (or iron(III) chloride) catalyst
   (b) C = Br_2; D = HBr; iron(III) bromide catalyst (or aluminium bromide catalyst)
   (c) E = $C_6H_5NO_2$; F = H_2O; concentrated nitric acid; concentrated sulfuric acid catalyst; approximately 50 °C

Phenol

1 See text.
2 (a) $C_6H_5OH + KOH \rightarrow C_6H_5O^-K^+ + H_2O$
   (b) $C_6H_5OH + Na \rightarrow C_6H_5O^-Na^+ + ½H_2$
   (c) $C_6H_5OH + 3Cl_2 \rightarrow C_6H_2Cl_3OH + 3HCl$
   (d) $C_6H_5OH + Li \rightarrow C_6H_5O^-Li^+ + ½H_2$
3 See text.
4 With phenol the reaction takes place with bromine water whilst benzene requires undiluted bromine liquid. Phenol does not require a catalyst to react whilst benzene requires a halogen carrier (e.g. $AlCl_3$).
5 See text.

Carbonyl compounds

1 (a) butanal
   (b) 2-methylpropanal
   (c) phenylmethanal
   (d) 2-phenylethanal
2 (a) propanone
   (b) pentan-3-one
   (c) phenylpropanone
   (d) 4-phenylbutan-2-one
3 (a) (i) CH_3CH_2OH
      (ii) $CH_3CH(OH)CH_2CH_3$
   (b) (i)   1-phenylethanol, $CH_3CH(C_6H_5)OH$
      (ii) 2-methylpropan-1-ol, $CH_3CH(CH_3)CH_2OH$
4 Add 2,4-dinitrophenylhydrazine solution to the compound. If a yellow/orange crystalline precipitate is obtained, then it is a carbonyl compound (either aldehyde or ketone).
   Test the compound with Tollens' reagent – if a silver mirror is obtained then it is an aldehyde; if no silver mirror then it is a ketone.
   Add 2,4-dinitrophenylhydrazine solution to the compound to form the 2,4-dinitrophenylhydrazone; filter off precipitate, purify, wash and dry.
   Test melting point of the 2,4-dinitrophenylhydrazone and match to data for known 2,4-dinitrophenylhydrazones of aldehydes or ketones.

Carboxylic acids

1 (a) CH_3CH_2COOH
   (b) $CH_3CH(CH_3)COOH$
   (c) $C_6H_5CH_2COOH$
2 See text.
3 (a) $CH_3CH_2COOH + NaOH \rightarrow CH_3CH_2COO^-Na^+ + H_2O$
   (b) $2CH_3CHClCOOH + Na_2CO_3 \rightarrow$
      $2CH_3CHClCOO^-Na^+ + H_2O + CO_2$
   (c) $C_6H_5COOH + Na \rightarrow C_6H_5COO^-Na^+ + ½ H_2$
   (d) $2CH_3CH(CH_3)COOH + Mg \rightarrow (CH_3CH(CH_3)COO^-)_2Mg + H_2$
   (e) $C_6H_5COOH + NaOH \rightarrow C_6H_5COO^-Na^+ + H_2O$
4 (a) $CH_3CH_2COOH + CH_3OH \rightleftharpoons CH_3CH_2COOCH_3 + H_2O$
   (b) $(CH_3CH_2CO)_2O + CH_3OH \rightarrow$
      $CH_3CH_2COOCH_3 + CH_3CH_2COOH$
   (c) $CH_3CHClCOOH + CH_3CH_2CH_2OH$
      $\rightleftharpoons CH_3CHClCOOCH_2CH_2CH_3 + H_2O$
   (d) $(CH_3CO)_2O + CH_3CH_2CH_2OH \rightarrow$
      $CH_3COOCH_2CH_2CH_3 + CH_3COOH$
   (e) $(CH_3CO)_2O + C_6H_5CH_2OH \rightarrow CH_3COOCH_2C_6H_5 + CH_3COOH$

Esters

1 (a) CH_3CH_2COOH. CH_3COOCH_3, $HCOOC_2H_5$
   (b) $CH_3CH_2CH_2COOH$, $CH_3CH(CH_3)COOH$, $CH_3COOCH_2CH_3$,
      $CH_3CH_2COOCH_3$, $HCOOCH_2CH_2CH_3$, $HCOOCH(CH_3)CH_3$
2 (a) (i)   propanoic acid, CH_3CH_2COOH
      (ii) methanoic acid, HCOOH
      (iii) butanoic acid $CH_3CH_2CH_2COOH$
   (b) (i)   methyl propanoate
      (ii) methyl methanoate
      (iii) ethyl butanoate
3 See text.
4 (a) $CH_3(CH_2)_5CH=CH(CH_2)_7COOH$
   (b) $CH_3(CH_2)_4CH=CHCH_2CH=CH(CH_2)_7COOH$
5 See text.
6 (a) $CH_3COOC_2H_5 + H_2O \rightarrow CH_3COOH + C_2H_5OH$
   (b) $C_6H_5COOCH_3 + NaOH \rightarrow C_6H_5COO^-Na^+ + CH_3OH$
   (c) $CH_3CH_2COOC_6H_5 + H_2O \rightarrow CH_3CH_2COOH + C_6H_5OH$
   (d) $HCOOCH_2CH_2CH_3 + H_2O \rightarrow HCOOH + CH_3CH_2OH$

Answers to quick check questions

Nitrogen compounds

1 **(a)** methylamine
  (b) propylamine
  (c) 2-methylpropylamine
  (d) phenylamine
2 **(a) (i)** $CH_3NH_3^+ + OH^-$
     (ii) $C_6H_5NH_3^+ + OH^-$
  (b) They can **accept protons** because the lone-pair electrons on the nitrogen can form a dative covalent bond with the proton.
3 **(a)** $CH_3NH_2 + HCl \rightarrow CH_3NH_3^+ + Cl^-$
  (b) $CH_3CH_2CH_2NH_2 + HCl \rightarrow CH_3CH_2CH_2NH_3^+ + Cl^-$
     (or $CH_3CH_2CH_2NH_3^+ Cl^-$)
  (c) $C_6H_5NH_2 + HCl \rightarrow C_6H_5NH_3^+ + Cl^-$
     (or $C_6H_5NH_3^+Cl^-$)
  (d) $2CH_3NH_2 + H_2SO_4 \rightarrow 2CH_3NH_3^+ + SO_4^{2-}$
     (or $(CH_3NH_3^+)_2SO_4^{2-}$)
     (or $CH_3NH_2 + H_2SO_4 \rightarrow CH_3NH_3^+ HSO_4^-$)
  (e) $CH_3NH_2 + CH_3COOH \rightarrow CH_3NH_3^+ + CH_3COO^-$
     (or $CH_3NH_3^+ CH_3COO^-$)
4 **(a)**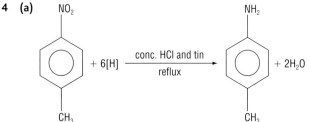
  (b)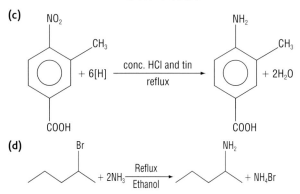
  (c)
  (d)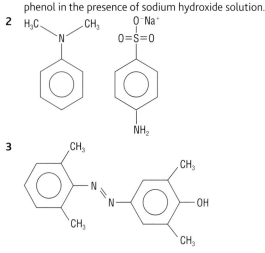

Phenylamine and azo compounds

1 React the phenylamine with a mixture of sodium nitrite (sodium nitrate(III)) and hydrochloric acid at 5°C.
  Take the benzenediazonium chloride formed and react it with phenol in the presence of sodium hydroxide solution.
2
3

Module 2 – Polymers and synthesis

Amino acids

1 **(a) (i)** $CH_2(NH_2)COOH + HCl \rightarrow CH_2(N^+H_3)COOH + Cl^-$
     (ii) $CH_2(NH_2)COOH + NaOH \rightarrow CH_2(NH_2)COO^-Na^+ + H_2O$
  (b) In **(i)** the nitrogen on the NH_2 group is accepting a proton and in **(ii)** the –COOH group donates a proton.

Amino acids, proteins and particles

1 **(a)** $^+H_3N-CH_2-COO^-$
  (b) $^+H_3N-CH(CH_3)-COO^-$
  (c) $^+H_3N-CH(C_6H_5)-COO^-$
2 **(a) (i)** $^+H_3N-CH_2-COOH$
     (ii) $^+H_3N-CH_2-COO^-$
     (iii) $H_2N-CH_2-COO^-$
  (b) (i) $^+H_3N-CH(CH_3)-COOH$
     (ii) $H_2N-CH(C_6H_5)-COO^-$
3 Ionic bonds are formed between neighbouring zwitterions. These are stronger than intermolecular forces such as hydrogen bonds, dipole–dipole or van der Waals' forces.
4 **(a) (i)** $H_2NCH_2CONHCH(CH_3)COOH$ and $H_2NCH(CH_3)CONHCH_2COOH$
     (ii) $H_2NCH_2CONHCH(C_6H_5)COOH$ and $H_2NCH(C_6H_5)CONHCH_2COOH$
     (iii) $H_2NCH_2CONHCH_2COOH$
  (b) (i) $H_2NCH(CH_3)COOH$ and $H_2NCH(CH_2C_6H_5)COOH$
     (ii) $H_2NCH(CH_3)COOH$ and $H_2N-CH(CH(CH_3)_2)COOH$

Stereoisomerism

1 **(a) (i)**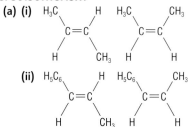
     (ii)
  (b) But-1-ene has two hydrogens on one of the carbons of the C=C bond. For E/Z isomerism both carbons of the C=C bond have to have two different atoms or groups attached.
  (c) Pent-2-ene has two hydrogens, one on each of the two carbons of the C=C bond. This means that it exhibits cis–trans isomerism, which is a special example of E/Z isomerism. The other compound lacks this arrangement so cannot exhibit cis–trans isomerism.
2 Enzymes are chiral catalysts and will produce chiral compounds. Therefore each enzyme for a particular reaction will produce only one optical isomer.
3 **(a)**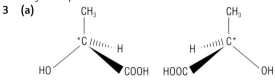
  (b) It does not have a carbon with four different groups or atoms attached.
  (c)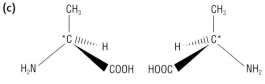

4

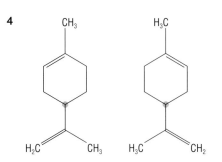

Polymerisation

1 (a)

(b)

(c)

(d)

2 (a)

(b)

3 (a) See text.
(b) See text.

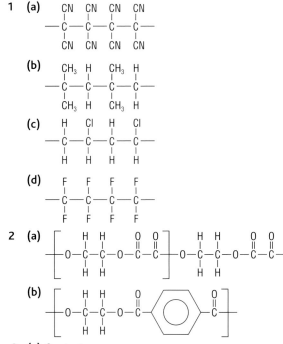

Condensation polymers

1 (a)

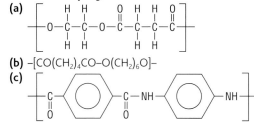

(b) $-[CO(CH_2)_4CO-O(CH_2)_6O]-$

(c)

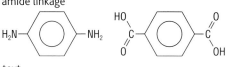

2 (a) $HOOCCH_2CH_2COOH$ and $HOCH_2CH_2OH$ ester linkage
(b) $HOOC(CH_2)_4COOH$ and $HO(CH_2)_6OH$ ester linkage
(c) amide linkage

3 See text.
4 (a) See text.
(b) See text.

5

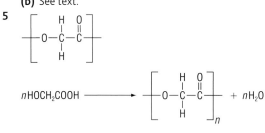

Organic synthesis including chiral synthesis

1. Because natural systems use enzymes as the catalysts; enzymes are chiral catalysts and will produce just one optical isomer.
2. See text.
3. **(a)** • boil the ester ethyl ethanoate under reflux with aqueous acid (e.g. HCl).
   $CH_3COOC_2H_5 + H_2O \rightleftharpoons CH_3COOH + C_2H_5OH$
   • The ethanol produced is then boiled under reflux with excess acidified potassium dichromate(VI) to give ethanoic acid.
   $C_2H_5OH + 2[O] \rightarrow CH_3COOH + H_2O$
   (b) • The propanal is reacted with sodium borohydride in water to produce propan-1-ol.
   $CH_3CH_2CHO + 2[H] \rightarrow CH_3CH_2CH_2OH$
   • The propan-1-ol is then refluxed with ethanoic acid in the presence of an acid catalyst (e.g. H_2SO_4) to give the ester. Alternatively the propan-1-ol is reacted with ethanoic anhydride in the cold.
   $CH_3CH_2CH_2OH + CH_3COOH \rightleftharpoons CH_3COOC_3H_7 + H_2O$
   (or $CH_3CH_2CH_2OH + (CH_3CO)_2O \rightarrow$
   $CH_3COOC_3H_7 + CH_3COOH$)

4. **(a) A** = ethanol; C_2H_5OH
   B = ethanoic acid; CH_3COOH
   C = ethyl ethanoate; $CH_3COOC_2H_5$
   (b)

   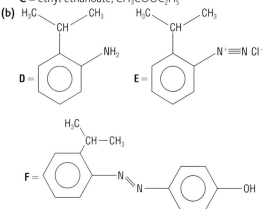

5. **(a)** ketone; alkene; secondary alcohol
   (b) (i) alkene
   (ii) secondary alcohol
   (iii) ketone

Module 3 – Analysis

Chromatography

1. See text.
2. R_f of P = 6.3/10 = 0.63; R_f of Q = 5.6/7 = 0.80; R_f of R = 6.5/9 = 0.72;
   R_f of unknown = 4.8/6 = 0.80; Unknown = Q.
3. **(a)** The order in which they leave the column would be reversed.
   (b) The ethanol is the more polar and therefore would be adsorbed more strongly onto the stationary phase on the column and would therefore be retained longer.

C13 NMR

1. **(a)** One peak at 135–160 ppm
   (b) 2 peaks, one at δ = 20–30 ppm (the methyl carbon) and one at 190–220 ppm (the carbonyl carbon)
   (c) Two peaks (one the >COOH carbon at 160–185 ppm); CH_3 carbon at 10–50 ppm
   (d) One peak at 115–140 ppm for C=C carbon.

2 P is for $CH_3COOCH_2CH_3$ because it has four peaks and there are four chemically different carbons in the molecule.
Q is for $CH_3CH(CH_3)COOH$ because there are three peaks and there are three chemically different carbons present in the molecule.

Proton NMR

1 (a) There are four types of chemically distinct protons.
(b) There are two chemically different protons on an adjacent carbon.

2 (a)

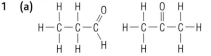

(b) 1,2-dichloroethane
(c) There is just one peak and therefore just one type of proton. The other isomer has two types of chemically distinct protons.

Identifying compounds using NMR spectroscopy

1 (a)

(b) The isomer is propanal. It would give three peaks because of the three different types of proton environment present. The peak at $\delta = 9.7$ ppm is due to the aldehyde proton. This peak would be split into a triplet because of the two chemically different protons (the $-CH_2-$ protons) on the adjacent carbon.

2 The structure of ethanal is shown

There will be two peaks because there are two types of chemically different proton: the methyl protons (1–3) and the CHO proton (proton 4).
The peaks will be at $\delta = 2.1$ ppm for the methyl protons (because they are adjacent to a carbonyl group) and at $\delta = 9.7$ ppm for the aldehyde proton.
The peak for protons 1–3 will be split into a doublet because of the single chemically different proton on the adjacent carbonyl carbon – thus obeying the $n + 1$ rule.
The peak for the aldehyde proton will be split into a quartet or quadruplet because of the three chemically different methyl protons on the adjacent carbon ($n + 1 = 3 + 1 = 4$ peaks).

3 (a)

(b) Two types of proton – the COOH proton and the CH_3 protons.
(c) The carboxyl (CO**OH**) proton at $\delta = \sim11.7$ ppm and the $-CH_3$ protons at $\delta = \sim2$ ppm.
(d) The peak at $\delta = \sim11.7$ ppm will disappear because the carboxyl proton will exchange with the D_2O to give $-COOD$, which will not produce a peak in this spectrum.

Combined analytical techniques

1 The peak at $m/z =15$ is due to the formation of the CH_3^+ ion when the molecular ion fragments.
$CH_3CHO^+ \rightarrow CH_3^+ + CHO\bullet$
The peak at $m/z = 29$ is due to the CHO^+ ion.
$CH_3CHO^+ \rightarrow CH_3\bullet + CHO^+$

---

2 (a) A sharp absorption at 1680–1750 cm^{-1} due to the presence of the $>C=O$ bond.
(b) A has five peaks; B has three peaks; C has four peaks.
(c) It must be C (3-methylbutan-2-one) since this has four peaks on the ^{13}C NMR because of having four chemically different carbons present. It has a relative molecular mass of 86 ($C_5H_{10}O$). The peak at $m/z = 43$ on the mass spectrum is due to either the CH_3CO^+ ion or the $(CH_3)_2CH^+$ ion.
$CH_3COCH(CH_3)CH_3^+ \rightarrow (CH_3)_2CH^+ + CH_3CO\bullet$
$CH_3COCH(CH_3)CH_3^+ \rightarrow (CH_3)_2CH\bullet + CH_3CO^+$

The proton NMR for C would give three peaks because there are three types of chemically different proton.
The peak at $\delta = 2.2$ ppm is due to the three methyl protons (H_a). It is a singlet because there are no chemically different protons on the adjacent carbon.
The heptuplet at $\delta = 2.6$ ppm is due to the single proton H_b. It is a heptuplet because there are *six* chemically different protons (H_c) on the adjacent carbons.
The doublet at $\delta = 1.3$ ppm is due to the six protons H_c on the two methyl groups that are chemically identical. The peak is a doublet because of the single chemically different proton H_b on the adjacent carbon.

Unit 2 Equilibria, energetics and elements

Module 1 – Rates, equilibrium and pH

How fast? The rate equation

1 (a) First order with respect to NO
(b) Second order overall

2 $dm^3\ mol^{-1}\ s^{-1}$

3 (a) Rate = $k[A][B]^2$
(b) 3rd
(c) $dm^6\ mol^{-2}\ s^{-1}$
(d) 1st
(e) 1st
(f) zero
(g) 2nd
(h) $dm^3\ mol^{-1}\ s^{-1}$
(i) Rate = $[A][B]^2$
(j) 3rd
(k) $dm^6\ mol^{-2}\ s^{-1}$
(l) Rate = $k[C]^2$
(m) 2nd
(n) $dm^3\ mol^{-1}\ s^{-1}$

Calculating k, the effect of T and reaction mechanisms

1 $3.00 \times 10^{-4}\ s^{-1}$

2 The rate-determining step is the reaction between a CH_3COCH_3 molecule and an $H^+(aq)$ ion.

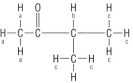

How fast? Concentration–time graphs
1 Yes, it has a constant half-life.
2 (a) $2H_2O_2 \rightarrow 2H_2O + O_2$
  (b) Rate = $k[H_2O_2]$
  (c) Graph plotted where the concentration decreases by half over a constant period of time.

How fast? Rate–concentration graphs
1 Measure the initial rate of a reaction in several experiments in which the reactants have different concentrations. For graph see text.
2 A graph of rate against concentration is a straight line through origin, so this is a first-order reaction.
3 (a) 1st order with respect to A;   Zero order with respect to B.
  (b)

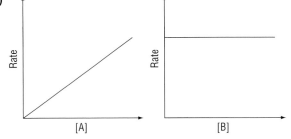

How far? The equilibrium law and K_c
1 (a) $K_c = \dfrac{[PCl_3][Cl_2]}{[PCl_5]}$ units = mol dm^{-3}

  (b) $K_c = \dfrac{[C_2H_5OH]}{[C_2H_4][H_2O]}$ units = dm^3 mol^{-1}

  (c) $K_c = \dfrac{[HBr]^2}{[H_2][Br_2]}$ units = no units

2 (a) Decreases   (b) No effect
  (c) Increases   (d) Increases
  (e) Increases   (f) No effect
  (g) Decreases   (h) Decreases
  (i) No effect   (j) No effect
  (k) Decreases   (l) Decreases

How far? How to calculate the value of K_c

1
$$CH_3COOC_2H_5 + H_2O \rightleftharpoons CH_3COOH + C_2H_5OH$$

At start 0.204 mol 0.645 mol 0 0
At equilibrium 0.114 mol

Since 1 mol of C_2H_5OH is formed along with every 1 mol of CH_3COOH formed, at equilibrium amount of C_2H_5OH is also 0.114 mol.

For every mol of CH_3COOH formed 1 mol of $CH_3COOC_2H_5$ must have reacted

So at equilibrium amount of $CH_3COOC_2H_5$ = 0.204 – 0.114 mol = 0.0900 mol

Similarly at equilibrium amount of H_2O = 0.645 – 0.114 = 0.531 mol

$K_c = \dfrac{[CH_3COOH][C_2H_5OH]}{[CH_3COOC_2H_5][H_2O]} = \dfrac{(0.114/V)(0.114/V)}{(0.0900/V)(0.531/V)}$

= 0.272 (no units)

2
	H_2	+	I_2	$\rightleftharpoons$	2HI
At start	1 mol		1 mol		0
At equilibrium	1.00 – 0.80		1.00 – 0.80		1.60 mol
	0.20 mol		0.20 mol		1.60 mol

$K_c = \dfrac{[HI]^2}{[H_2][I_2]} = \dfrac{(1.6/V)^2}{(0.2/V)(0.2/V)} = 64.0$ (no units)

Acids and bases
1 (a) <u>H</u>ClO$_4$ and ring ClO$_4^-$
  (b) <u>H</u>Br and ring Br$^-$
  (c) <u>H</u>$_2$O and ring OH$^-$
2 (a) Hydronium ion
  (b) In **1(b)** the water accepts a proton (H$^+$ ion) from HBr and is therefore a base. In **1(c)** it donates a proton to the CH_3NH_2 and therefore acts as an acid. Because it can act as both an acid and a base it is amphoteric.
3 (a) $HClO_4 + H_2O \rightleftharpoons ClO_4^- + H_3O^+$
     $HNO_3 + H_2O \rightleftharpoons NO_3^- + H_3O^+$
  (b) The $HClO_4$ is the stronger proton donor because it is the acid in its reaction with HNO_3.
4 (a) $Mg(s) + 2H^+(aq) \rightarrow Mg^{2+}(aq) + H_2(g)$
  (b) $CuCO_3(s) + 2H^+(aq) \rightarrow Cu^{2+}(aq) + H_2O(l) + CO_2(g)$
  (c) $CaO(s) + 2H^+(aq) \rightarrow Ca^{2+}(aq) + H_2O(l)$
  (d) $H^+(aq) + OH^-(aq) \rightarrow H_2O(l)$

Acids and bases in aqueous solution
1 (a) 2
  (b) 8.7
  (c) 3.52
2 (a) 3.16×10^{-3} mol dm^{-3}
  (b) 3.98×10^{-8} mol dm^{-3}
  (c) 3.16×10^{-11} mol dm^{-3}
  (d) 2.51×10^{-14} mol dm^{-3}
3 (a) (i) 2
      (ii) 4.52
  (b) (i) 12
      (ii) 9.48

The chemistry of weak acids
1 (a) pK_a (HA) = 2.54, pK_a (HB) = 5.35
  (b) HA is the stronger acid because it has the lower pK_a (or conversely the higher K_a).
  (c) $HA + HB \rightleftharpoons A^- + H_2B^+$
2 $[H^+(aq)] = 10^{-pH} = 10^{-3.41} = 3.89 \times 10^{-4}$
  $K_a = [H^+(aq)]^2/0.01 = (3.89 \times 10^{-4})^2/0.01 = 1.51 \times 10^{-5}$ mol dm^{-3}
3 (a) $K_a = 10^{-pK_a} = 10^{-9.9} = 1.26 \times 10^{-10}$ mol dm^{-3}
  (b) $[H^+(aq)] = \sqrt{(1.26 \times 10^{-10} \times 0.001)} = 3.55 \times 10^{-7}$
     pH = $-\log_{10}[H^+(aq)]$ = 6.45

Buffers – how they work and their pH
1 See text.
2 (a) [HF] = 100/150 $\times$ 0.03 = 0.02 mol dm^{-3}
  (b) [NaF] = 50/150 $\times$ 0.03 = 0.01 mol dm^{-3}
  (c) $[H^+(aq)] = K_a \times$ [HF]/[NaF] = $5.6 \times 10^{-4} \times 0.02/0.01$
              = 1.12×10^{-3} mol dm^{-3}
     pH = $-\log_{10}[H^+(aq)] = -\log_{10}1.12 \times 10^{-3}$ = 2.95
     $\left(\text{or use pH} = pK_a + \log_{10}\dfrac{[\text{salt}]}{[\text{acid}]}\right)$
  (d) The equilibrium for the acid is
     $HF(aq) \rightleftharpoons H^+(aq) + F^-(aq)$
     If H$^+$(aq) ions are added to disturb the equilibrium, they combine with the conjugate base, the F$^-$(aq) ions from the NaF, to give the largely undissociated HF acid. This removes them from the solution and the pH remains virtually constant. If OH$^-$(aq) ions are added they are neutralised by the H$^+$(aq) ions. This disturbs the equilibrium and more HF dissociates so that the value of K_a is restored, along with the concentration of H$^+$(aq) ions, and the pH returns to its original value.
     See text.
3 See text.

Indicators, acid–base titration curves and neutralisation

1

Indicator	(a) Colour at pH 1.0	(b) Colour at pH 7.0	(c) Colour at pH 10.0
Congo red	violet	red	red
Methyl red	red	yellow	yellow
Thymol blue	yellow	yellow	blue
Thymolphthalein	colourless	colourless	pale

2 (a) Any of the four can be used.
   (b) Thymol blue or thymolphthalein.
   (c) Congo red or methyl red.
   (d) No indicator can be used.
       Total mass of water = 200 g;
       Enthalpy change = m × shc × Δt = 200 × 4.18 × 13.8 J
       $\qquad = -11\,536$ J $= -11.54$ kJ
3 The amount in mol of H^+ ions = 2.00 × 0.100 = 0.200 mol
   Enthalpy of neutralisation = – 11.54/0.200 kJ mol^{-1}
   $\qquad = -57.7$ kJ mol^{-1}

Module 2 – Energy

Lattice enthalpy

1 (a) NaCl has the more exothermic lattice enthalpy, because Na^+ is smaller than Cs^+ so has a higher charge density. Therefore it has a higher electrostatic attraction to Cl^- in the lattice.
   (b) NaF because the F^- ion is smaller and therefore has the greater charge density and so attracts the Na^+ ion more strongly, hence a more exothermic lattice enthalpy.
2 Mg^{2+} (g) + $2Cl^-$(g) → $MgCl_2$(s)
3 The Mg^{2+} ion and the oxide ion, O^{2-}, are smaller and have a greater charge than both the Na^+ and the Cl^- ion.
   Therefore the charge density on both the Mg^{2+} and the O^{2-} ions is much greater. They attract each other more strongly and therefore MgO has a much more exothermic lattice energy than NaCl.

The Born–Haber cycle

1 Li$^+$(g) + e$^-$ + F(g) ——————

   ΔH_{at}(F) ———————————— 1st electron affinity of F

   Li$^+$(g) + e$^-$ + ½ F$_2$(g)   Li$^+$(g) + F$^-$(g)

   1st I.E. of Li

   Li$^+$(g) + e$^-$ + ½ F$_2$(g)

   ΔH_{at}(Li) ———————————— Lattice enthalpy of LiF

   Li(s) + ½ F$_2$(g)

   ΔH_f(LiCl)

   ——————————— Li$^+$F$^-$(s)

2 Lattice enthalpy = – 1046 kJ mol^{-1}
3 (i)   Li(s) + ½ F$_2$(g) → LiF(s)
   (ii) Li(s) → Li(g)
   (iii) Li(g) → Li$^+$(g) + e$^-$
   (iv) ½ F$_2$(g) → F(g)
   (v)   F(g) + e$^-$ → F$^-$(g)

Things to know about Born–Haber cycles and hydration enthalpies

1 ΔH_f (CaCl$_2$) = (+178) + (+590) + (+1145) + 2 × (121.7) + 2 × (–348.8)
   $\qquad$ + (–2258)
   $\qquad = -799$ kJ mol^{-1}
2 Li^+. It is a smaller ion and therefore its charge density is greater. This means that it will attract the dipoles on water molecules more strongly and therefore form stronger ion–dipole attractions and hence the hydration enthalpy will be stronger.
3 Ca^{2+}: the greater charge means that it has a higher charge density and therefore has a greater attraction for the dipoles on the water molecules.
4 $\Delta H_{solution} = -\Delta H_{lattice} + (\Delta H_{hydration}$ of Na^+ and $Cl^-)$
   $\Delta H_{hydration}(Na^+) = \Delta H_{solution} + \Delta H_{lattice} - \Delta H_{hydration}$ (Cl$^-$)
   $\qquad = +1\,-770 + 384 = -385$ kJ mol^{-1}

Enthalpy and entropy

1 (a) $\Delta G = \Delta H - T\Delta S = 100 - 298 \times 121 \times 10^{-3}$ kJ mol^{-1}
       $\qquad = +63.9$ kJ mol^{-1}
       This is positive and therefore the reaction *will not proceed spontaneously* at 298 K.
   (b) $\Delta G = \Delta H - T\Delta S = -101$ kJ mol^{-1}
       This is negative and therefore the reaction *will proceed spontaneously* at 298 K.

Redox reactions and electrode potentials

1 Gain of electrons and a decrease in oxidation number or state.
2 The I^- ion has lost an electron and its oxidation state has increased from –1 to 0.
3

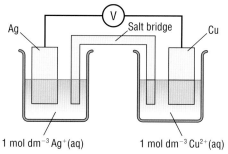

1 mol dm^{-3} Ag$^+$(aq)   1 mol dm^{-3} Cu^{2+}(aq)

The salt bridge completes the circuit and maintains the ionic balance in the half-cells.
4 (a) Br$_2$ + 2e$^-$ ⇌ **2Br$^-$**
   (b) **Fe^{2+}** ⇌ Fe^{3+} + e$^-$
   (c) H$_2$ ⇌ 2e$^-$ + **2H$^+$**
   (d) Mn^{3+} + **e$^-$** ⇌ Mn^{2+}
5 Because the metal is in contact with an aqueous solution of its ions.

Standard electrode potentials

1 See text.
2 See text.
3 The Zn^{2+}/Zn would form the negative terminal and the Cl_2/Cl^- would form the positive terminal.
4 (a) E_{cell} = 1.36 –(–2.37)= 3.73 V
   (b) E_{cell} = 0.77–(–0.76) = 1.53 V
   (c) E_{cell} = 1.36 – 0.77 = 0.59 V
5 The two half-cell reactions involved are:
   2e$^-$ + Zn^{2+}(aq) ⇌ Zn (s) $E^{\ominus} = -0.76$ V
   e$^-$ + Fe^{3+}(aq) ⇌ Fe^{2+}(aq) $E^{\ominus} = +0.77$ V
   No! For the reaction to take place the Fe^{2+} ion has to lose an electron. The standard electrode potential for the second reaction is more positive than for the first reaction and therefore it cannot lose electrons. It must accept electrons and it cannot proceed to the right and the top reaction cannot proceed to the right.

6 The equilibrium $2e^- + Cl_2(g) \rightleftharpoons 2Cl^-$ would shift to the right, making the electrode potential more positive. This would increase the difference between the electrode potentials, so E_{cell} would increase.

Fuel cells, fuel cell vehicles and the hydrogen economy

1 See text.

2 (a) $2e^- + 2H^+(aq) \rightleftharpoons H_2(g)$
(b) $2e^- + \frac{1}{2}O_2(g) + 2H^+(aq) \rightleftharpoons H_2O(l)$
(c) $6e^- + 6H^+(aq) + CO_2(g) \rightleftharpoons CH_3OH(l) + H_2O(l)$

3 See text.

4 See text.

5 See text.

Module 3 – Transition elements

Transition elements – electron configurations

1 See text.

2 Mn = [Ar] $3d^5 4s^2$ NOTE: [Ar] = $1s^2 2s^2 2p^6 3s^2 3p^6$
Cr = [Ar] $3d^5 4s^1$

3 Mn^{2+} = [Ar] $3d^5 4s^0$
Cr^{3+} = [Ar] $3d^3 4s^0$

4 The Cu^+ ion has the electron configuration [Ar] $3d^{10}$. Therefore it does not have a partially filled d-shell and so it is not coloured. The Cu^{2+} ion has the electron configuration [Ar] $3d^9$. Therefore it has a partially filled d-shell and is coloured.

Transition elements – oxidation states, catalytic behaviour and the hydroxides

1 See text.

2 See text.

3 (a) Fe^{3+} (iron(III))
$Fe^{3+}(aq) + 3OH^-(aq) \rightarrow Fe(OH)_3(s)$
(b) (i) yellow-orange
(ii) Fe^{2+} (iron(II))
(iii) green precipitate

Complex ions

1 A **ligand** is an ion or molecule with lone-pair electrons that are used to form a coordinate (dative covalent) bond to a central metal ion in a complex. In this complex the ligands are the water molecules.
A **complex** is a central metal ion (or sometimes an atom) surrounded by coordinate bonded (dative covalent) ligands. Here the central metal ion is the Cu^{2+} ion.
Octahedron is the shape of the complex and other complexes where six coordinate bonds are formed between the ligand(s) and the central metal ion.
Coordination number is the number of coordinate bonds formed by the ion with the ligand(s).

2 (a)   (b)

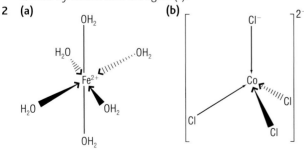

3 A bidentate ligand is one that has two pairs of electrons that it uses in coordinate bonding with the central metal ion in a complex. $H_2NCH_2CH_2NH_2$ has two pairs of electrons it uses – one on each nitrogen atom.

4

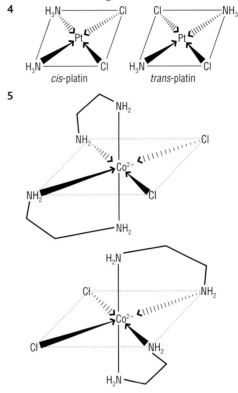

cis-platin   *trans*-platin

5

Ligand substitution

1 (a) The H_2O molecules in the $[Cu(H_2O)_6]^{2+}$ ion are replaced by NH_3 molecules.
(b) $K_{stab} = \dfrac{[Cu(NH_3)_4(H_2O)_2^{2+}]}{[Cu(H_2O)_6^{2+}][NH_3]^4}$ dm^{12} mol^{-4}

2 See text.

3 See text.

Redox reactions and calculations

1 Equation is
$2Fe^{2+}(aq) + H_2O_2(aq) + 2H^+(aq) \rightarrow 2Fe^{3+}(aq) + 2H_2O(l)$

2 (a) $[Fe^{2+}]$ = 5.56/(55.8 + 32.1 + 64 + (7 × 78))
= 0.0200 mol dm^{-3}
(b) Amount in mol of Fe^{2+} = 0.025 × 0.0200 = 5.00 × 10^{-4} mol.
Amount in mol of MnO_4^- = 1/5 × 5.00 × 10^{-4} mol
= 1.00 × 10^{-4} mol.
$[MnO_4^-]$ = 1.00 × 10^{-4}/0.02 = 0.0050 mol dm^{-3}.

3 (a) $2Cu^{2+}(aq) + 4I^-(aq) \rightarrow 2CuI(s) + I_2(aq)$
$I_2(aq) + 2S_2O_3^{2-}(aq) \rightarrow 2I^-(aq) + S_4O_6^{2-}(aq)$
(b) Amount in mol of $S_2O_3^{2-}$ = 0.100 × 0.0225 = 2.25 × 10^{-3} mol.
Amount in mol of Cu^{2+} = amount in mol of $S_2O_3^{2-}$
= 2.25 × 10^{-3} mol.
$[Cu^{2+}]$ = 2.25 × 10^{-3}/0.025 = 0.09 mol dm^{-3}.
Therefore amount in mol in 250 cm^3 = 0.25 × 0.09 mol
= 0.0225 mol.
Mass of copper = 0.0225 × 63.5 = 1.43 g.
Percentage of copper in brass = 1.43/2.30 × 100% = 65%.

Answers to end-of-unit questions

Unit 1 Rings, polymers and analysis

1 (a) (i) Let there be 100 g of the hydrocarbon

Element	%	Mass in 100 g	No. of moles	Relative no. of atoms
Carbon	90.56	90.56 g	90.56/12 = 7.55	4
Hydrogen	9.44	9.44 g	9.44/1 = 9.44	5 (because 9.44/7.55 = 1.25 = 5/4)

[1]

(ii) The empirical formula mass = (4 × 12) + (5 × 1) = 53
Relative molecular mass = 106 = 2 × 53
Therefore molecular formula = 2 × empirical formula
= 2 × C_4H_5 = C_8H_{10} [1]

(iii) [1] for each correct structure

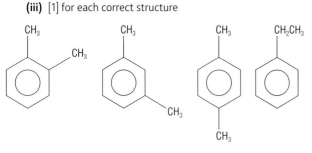

(b) (i) **A** must be the hydrocarbon with the methyl groups at the 1 and 4 positions. The product can only have one identity. The others give monochloro-products which have the chlorine in more than one possible position. (Draw them to prove it to yourself.)   [1]

(ii)

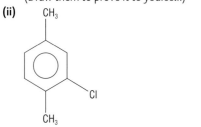

[1]

2 (a) To form optical isomers, a compound should possess a chiral carbon (one with four different groups attached to it). Glycine does not possess this property because it has two hydrogens attached to the central carbon atom.

(b) (i) $H_2NCH_2COOH + HCl \rightarrow H_3N^+CH_2COOH + Cl^-$
(or $Cl^- H_3N^+CH_2COOH$)   [1]

(ii) $H_2NCH(CH_3)COOH + NaOH \rightarrow$
$H_2NCH(CH_3)COO^-Na^+ + H_2O$ [1]

(c) (i) $H_3N^+CH_2COO^-$ [1]

(ii)

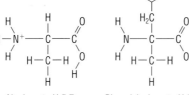

Alanine at pH 5.7   Phenylalanine at pH 5.7

You are not asked for an explanation but the reasoning is that pH 5.75 is below the isoelectric point of alanine, so H^+ ions can be added to the zwitterion giving a protonated carboxylate group.
pH 5.75 is above the isoelectric point of phenylalanine, so H^+ ions are removed giving a de-protonated zwitterion.

(d)

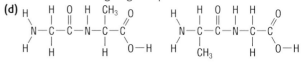

for each correct –NHCO– bond [1] mark and [1] mark for correct structure for rest of each molecule   [1]

3 (a) $CH_3CH_2CH_2CH_2CHO$, $CH_3CH(CH_3)CH_2CHO$,
$CH_3CH_2CH(CH_3)CHO$, $CH_3C(CH_3)_2CHO$,
$CH_3CH_2COCH_2CH_3$, $CH_3COCH_2CH_2CH_3$,
$CH_3COCH(CH_3)CH_3$ [1] mark each, [7] in total

(b) The substance is a ketone because it does not have a peak at δ = 9.7 ppm on the proton NMR.   [1]
Carbon-13 NMR shows that it has only three types of carbon (only three peaks).   [1]
Therefore it must be $CH_3CH_2COCH_2CH_3$.   [1]
This is supported by the fact that there are only two peaks for the proton NMR and this has only two types of chemically different proton.   [1]
Peak at δ = 2.5 ppm is produced by the CH_2 protons and is a quartet because of the three chemically different protons on adjacent carbon (the CH_3 protons).   [1]
Peak at δ = 1.2 ppm is produced by the CH_3 protons and is a triplet because of the two chemically different protons on adjacent carbon (the $-CH_2-$ protons).   [1]

(c) The isomer would not give a silver mirror with Tollens' reagent because it is a ketone.   [1]
When a solution of 2,4-DNP is added it will give a yellow-orange precipitate.   [1]
This is filtered off, [1] washed and purified (recrystallised) then washed and dried again.   [1]
Melting point of resulting solid determined.   [1]
Compare this with literature/data-book to see which 2,4-DNP derivative has the same melting point.   [1]

(d)

[1] for dipole
[1] for curly arrow from hydride and from double bond onto oxygen
[1] for intermediate
[1] for curly arrows from oxygen to hydrogen on water

4 (a) (i) hexadecanoic acid   [1]
(ii) hexadec-9-enoic acid 16,1(9) [1] for name and [1] for numbers   [1]
(iii) octadeca-9,13-dienoic acid 18,2(9,13) [1] for name and [1] for numbers   [1]

(b) $CH_3(CH_2)_{14}COOH + NaOH \rightarrow CH_3(CH_2)_{14}COO^-Na^+ + H_2O$
[1] for reactants and [1] for products
$CH_3(CH_2)_{14}COOH + C_2H_5OH \rightleftharpoons CH_3(CH_2)_{14}COO\, C_2H_5 + H_2O$
[1] for reactants and [1] for products

(c) They lead to formation of 'bad' cholesterol.   [1]
This can lead to atherosclerosis and heart and circulation problems in the body.   [1]

(d) (i) [1] each for structure and name [1]

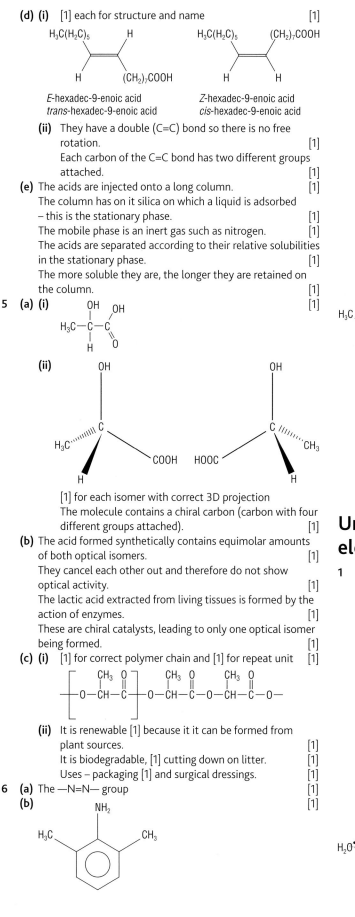

E-hexadec-9-enoic acid
trans-hexadec-9-enoic acid

Z-hexadec-9-enoic acid
cis-hexadec-9-enoic acid

(ii) They have a double (C=C) bond so there is no free rotation. [1]
Each carbon of the C=C bond has two different groups attached. [1]

(e) The acids are injected onto a long column. [1]
The column has on it silica on which a liquid is adsorbed – this is the stationary phase. [1]
The mobile phase is an inert gas such as nitrogen. [1]
The acids are separated according to their relative solubilities in the stationary phase. [1]
The more soluble they are, the longer they are retained on the column. [1]

5 (a) (i) [1]

(ii)

[1] for each isomer with correct 3D projection
The molecule contains a chiral carbon (carbon with four different groups attached). [1]

(b) The acid formed synthetically contains equimolar amounts of both optical isomers. [1]
They cancel each other out and therefore do not show optical activity. [1]
The lactic acid extracted from living tissues is formed by the action of enzymes. [1]
These are chiral catalysts, leading to only one optical isomer being formed. [1]

(c) (i) [1] for correct polymer chain and [1] for repeat unit [1]

(ii) It is renewable [1] because it it can be formed from plant sources. [1]
It is biodegradable, [1] cutting down on litter. [1]
Uses – packaging [1] and surgical dressings. [1]

6 (a) The —N=N— group [1]
(b) [1]

(c) STEP I: the amine is reacted with sodium nitrite [1] and HCl acid [1] at a temperature of 5 °C. [1]

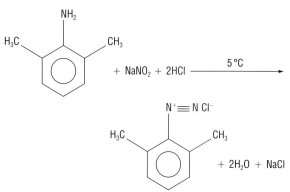

[1] for correct reactants and [1] for correct products
STEP II: The benzenediazonium chloride [1] formed is reacted with phenol in the presence of sodium hydroxide. [1]

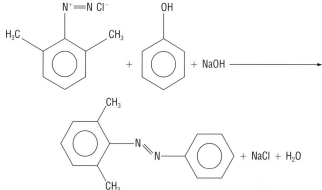

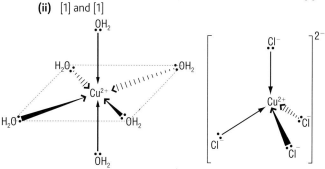

[1] for correct reactants and [1] for correct products

Unit 2 Equilibria, energetics and elements

1 (a) (i) I A covalent bond (sharing of a pair of electrons) where both electrons come from one atom. [1]
In the examples, the oxygen in the water, or the Cl^-, provides both electrons in each dative covalent bond to the metal ion. [1]

II A ligand is a molecule or ion, in this case the water molecule or the Cl^- ion, that donates an electron pair [1] to a central metal ion, in this case the Cu^{2+} ion. [1]

III The complexes in the example are formed by the Cu^{2+} ion accepting lone-pair electrons into its empty orbitals [1] from the (ligands) water or Cl^- ion. [1]

(ii) [1] and [1]

(b) (i) If changes are applied to a system in equilibrium, [1] the position of equilibrium will move to minimise [1] the effects of the change.

(ii) Solution changes to green and then to yellow [1] (these two must be in the correct order to obtain the mark). The equilibrium moves to the right [1] to reduce the concentration of Cl^-. [1]

(c) (i) dm^{12} mol^{-4} [1]

(ii) $4.17 \times 10^5 = [CuCl_4]^{2-}/1.17 \times 10^{-5} \times 0.8^4$ [1] $[CuCl_4]^{2-}$
= 2.0 mol dm^{-3} [1]

(iii) The concentration of water is very large [1] so is effectively constant. [1] Therefore the concentration of water can be incorporated into the equilibrium constant. [1] (2 out of these 3)

2 (a) (i) A strong acid is one that is completely dissociated in aqueous solution. [1]
Example HCl (aq) $\rightarrow$ H^+(aq) + Cl^-(aq) (can use HNO_3, H_2SO_4) [1]
A weak acid is one that is only partially dissociated in aqueous solution . [1]
Example $CH_3COOH \rightleftharpoons CH_3COO^-$(aq) + H^+(aq) [1]

(ii) A buffer solution resists change in pH (when small quantities of acid or alkali are added). [1]
Two equations: $CH_3COOH \rightleftharpoons CH_3COO^-$(aq) + H^+(aq)
and $CH_3COOH \rightarrow CH_3COO^-$(aq) + Na^+(aq) [1]
H^+ ions are mopped up by combining with the CH_3COO^- ion (conjugate base) [1]
to form the largely undissociated CH_3COOH (weak acid). [1]
OH^- ions are mopped up by reacting with the H^+ ions from the weak acid (neutralisation) to form water. [1]
This disturbs the equilibrium and more CH_3COOH dissociates to maintain the value of K_a. [1]

(b) (i) pH = 1.3 [1]

(ii) $[H^+] = 1 \times 10^{-14}/0.01$ [1]
pH = 12 [1]

(c) (i) $K_a = [H^+][C_6H_5COO-]/[C_6H_5COOH]$ [1]
$6.3 \times 10^{-5} = [H^+]^2/0.02$ [1]
pH = 2.95 [1]

(ii) Diagram to show:
shape of curve [1]
vertical section at 25 cm^3 [1]
graph starts at figure calculated in **(i)** or 2–4 and finishes at 12–14. [1]

(iii) Thymolphthalein [1]
The pH must change quickly with a small addition of alkali *or* pH change of indicator must be on vertical region of graph. [1]

3 (a) (i) $CH_3OH(l) + 1\frac{1}{2}O_2(g) \rightarrow CO_2(g) + 2H_2O(l)$ [1]

(ii) One mark for getting the enthalpy changes correct –239, 0 –394 + 2 × –286 used (use of balancing numbers) [1]
Then equating the changes to the enthalpy change of reaction (i.e. correct cycle used)
(–394 + 2 × –286) – (–239) [1]
–727 kJ mol^{-1} [1] (correct numerical answer; ignore units)

(iii) $\Delta S = (214 + 2 \times 70) - ((1.5 \times 103) + 127)$ [1]
= +727 J K mol^{-1} [1]

(b) $\Delta G = \Delta H - T\Delta S$ [1]
= –727 – (298 × 727/1000) [1]
= ~~–745 kJ mol⁻¹~~ ~ 945 kJ mol⁻¹ [1]
The value of ΔG is negative and therefore the change is spontaneous. [1]

(c) The activation energy is too high for the reaction to proceed at 298 K. [1]
Energy is required to give particles enough energy to react. [1]

(d) (i) Product is NO_2 (or N_2O_4). [1]
$O_2 + 2NO \rightarrow 2NO_2$

(ii) On the left-hand side the oxygen in O_2 has an oxidation state equal to 0. On the right-hand side it is –2. Therefore it has been reduced. [1]
The nitrogen goes from oxidation state +2 to +4 on the right-hand side. Therefore it has been oxidised since its oxidation state has increased. [1]

(iii) The order with respect to O_2 is first order, [1]
since, using experiments 2 and 4 for example, doubling $[O_2]$ leads to a doubling of the rate. [1]
The order with respect to NO is second order, [1]
since, using experiments 1 and 2, doubling [NO] causes the rate to quadruple. [1]

(iv) (I) Rate = $k[O_2][NO]^2$

(II) Using experiment 1,
$k = rate/[O_2][NO]^2$
$= 0.7 \times 10^{-4}/1 \times 10^{-2}$
$\times (1 \times 10^{-2})^2$ mol dm^{-3} s^{-1}/(mol $dm^{-3})^3$
= 70[1] dm^6 mol^{-2} [1]

4 (a) (i) The standard electrode potential ($E^\ominus$) of a half-cell is the e.m.f. of the half-cell using a standard hydrogen electrode as the reference electrode. [1]
All measurements are at 298 K and 100 kPa and all solutions are of 1.00 mol dm^{-3} concentration. [1]

(ii) Yes. The electrode potential of the Cu^+/Cu system is more negative than that of the Fe^{3+}/Fe^{2+} system. [1]
Therefore the bottom reaction can proceed to the left and lose electrons. [1]
The top reaction can gain electrons and proceed to the right. [1]

(b) (i) Any two from:
There is a brown gas formed; [1] this is NO_2 [1]
The solution turns blue [1] because Cu^{2+}(aq) is blue [1]
The copper dissolves; [1] Cu^{2+} is in solution [1]
The copper changes in oxidation state from 0 to +2 and is therefore oxidised [1]
The nitrogen changes in oxidation state from +5 (in NO_3^-) to +4 (in NO_2) and is therefore reduced. [1]

(ii) amount in mol of Cu^{2+} = amount in mol of $S_2O_3^{2-}$ [1]
amount in mol of $S_2O_3^{2-}$ = 0.100 × 0.0221 = 2.21 × 10⁻³ mol; concentration of Cu^{2+} [1]
$[Cu^{2+}$(aq)] = 2.21 × 10⁻³/ 0.025 = 8.84 × 10⁻² mol dm^{-3} [1]

(iii) amount in mol of Cu^{2+} in 250 cm^3 = 8.84 × 10⁻² × 0.25
= 2.21 × 10⁻² mol [1]
mass of copper present = 2.21 × 10⁻² × 63.5 [1] = 1.40g [1]

(iv) Percentage by mass of copper = 1.40/2.26 × 100% [1]
= ~~62.1~~% 61.9 % [1]

5 (a) The lattice enthalpy is the enthalpy change when 1 mol of solid ionic lattice [1]
is formed from its constituent gaseous ions. [1]

(b) (i) As the group is descended the lattice enthalpy becomes less exothermic. [1]
This is because the metal cations are larger, having lower charge density. [1]
The attraction between anions and cations decreases, giving a less exothermic lattice enthalpy. [1]

(ii) There is electrostatic attraction [1]
between the cations and the water molecules [1]
ion–dipole attractions [1]

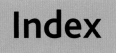

Index

Index